1984

ROBOT
TECHNOLOGY
SERIES

BOOK
TWO

ROBOTS

THE APPLICATION OF ROBOTS TO PRACTICAL WORK

BY DAVID M. OSBORNE

Midwest Sci=Tech
Publishers, Inc.

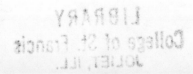

 Author David M. Osborne is Manager of Training and Service for Steelweld Robotic Systems, a unit of United Technologies. An alumnus of Wayne State University and The University of Michigan, Mr. Osborne also is an instructor at Lawrence Institute of Technology. His career in robotics began in 1968 with an attempt to develop a beer-retrieval robot. Over the years, he has been employed at automotive equipment, material handling, automation and robot companies serving in capacities from salesman to senior project engineer.

Dedicated to Timothy J. Osborne,
modelmaker and good scout. Also, a
special thanks to Jay Allen Dallaire,
Nancy M. Osborne and Anita Dallaire.

CONTENTS

ILLUSTRATIONS

FOREWORD

Learning to drive from a high school driving instructor is different from learning to race from an Indy 500 driver. Somewhat the same situation exists in learning about robots. It is possible to know how a robot works and what features are needed for an installation, yet not know the dynamics of a real application. Unquestionably, the best way to learn the practicality of robot applications is by experience, but experience is allowed mostly to those who already possess it. This volume is an attempt to train robot race drivers before they enter the track.

CHAPTER 1

WHY CONVERT TO ROBOTICS?

There are three main reasons why manufacturing jobs are converted to robotics.

1. Because of high technology, some jobs are better suited to robots than people.
2. Robotics provides greater safety for workers.
3. Robotics provides an economic advantage.

BETTER THAN PEOPLE

One task a robot often can do better than a human is visual inspection, for example, quality control checks at various production stages of parts being manufactured.   Modern vision equipment can analyze and measure complex parts better than the human eye.

The vision element is not a part of a basic robot. However, the element must be carried to the work piece or the work piece manipulated in front of it. This can be done quickly and efficiently by a robot.

Another task better suited for a robot is the cutting of fiberglass materials with a small cutting wheel. For best results, a steady motion is needed by the tool holder. With a human worker, there are variations in force and speed, sometimes caused by minor variations in the fiberglass material. A robot can manipulate the cutter smoothly along an exact path.

Still another job that can better be done by a robot is continuous bead gluing. The bead of adhesive is most effective if it is unbroken and of uniform height and shape. And a human cannot achieve the precision of a robot in guiding a glue nozzle at a controlled speed.

An area where a robot excels is in a job requiring extreme cleanliness. An example is the production of silicon wafers for use in semiconductor

1

chips. Contamination as small as bacteria is sufficient to cause failure in a semiconductor device. A human worker, even wearing a hospital gown and face mask, would present contamination problems, but a properly designed robot can manipulate components in a clean environment without incident.

A robot can also do a good job of assembling electronic components. A robot can manipulate parts with a specialized gripper exerting no more force than necessary, and each part can be placed with great precision. This careful handling is ideal for the delicate nature of silicon circuitry.

SAFETY FOR WORKERS

The second major reason for robot use is to protect human workers from dangerous environments. Press loading, for example, can be done by a robot. Even if no other economic or technological reasons existed for converting this task to robotics, the safety factor could justify the change.

Some manufacturing operations are considered to be sufficiently hazardous so that union representatives request the installation of robotic equipment. Other opportunities to use robots for worker safety involve working with toxic materials or in extreme temperature environments.

ECONOMIC ADVANTAGE

The third and most common reason for the use of robots is their economic advantage—their ability to produce manufactured goods more efficiently. This leads to cost savings and greater profits in several ways.

While an individual task assigned to a robot may show only a moderate return on investment, having that task robotized might speed up the entire manufacturing operation. Increasing productivity by 10 percent at one work station might mean increased productivity of 10 percent in the rest of the manufacturing process. The overall cost benefit achieved might dwarf the expense of the robotic conversion.

As an example of increased production, let's say that a manufacturer uses die cast machines capable of producing one part per minute and these machines are being used only at a rate of one part in 80 seconds because of the subsequent deburring operation. If the use of robots for deburring allows the die cast machines to function at their natural limit, overall production will increase. Without further changes in the manufacturing operation, the conversion of a job to a robotic one increases the efficiency of the whole process.

Another economic advantage can be greater quality. The value added to a piece of work in progress at a particular station may be small, but if a mistake is made by a human operator and a semifinished part becomes scrap, not only is value lost for that particular work station but also for all of the work done up to that time. A robot demonstrating a decreased scrap rate not only produces with greater efficiency at its own work station but also reduces the loss of work done at previous stations in the manufacturing process.

Robots have an economic advantage over their hard automation counterparts because they rapidly can be switched from one process to another. This is especially useful for short runs.

MUST BE A REASON

For a task to be converted to robotics, there must be a good reason. If there is no clear-cut advantage in the areas of technology, safety or economics, it may well be that the conversion will be unsuccessful. Perhaps the job will be more troublesome than it was previously.

EXPENSIVE TOY SYNDROME

There is always the chance that once a robot is considered for a particular work task, those assigned to the project may be caught up in the robot mystique and ignore the underlying rationale for the conversion. Once the ball is rolling, it is hard to stop. This might be termed the *Expensive Toy Syndrome*.

Here's a typical example: An automotive company decides to try a particular robot. A plant is selected for trial based on its lack of robots or its proximity to upper management. Orders come from top executives that a place must be found for a robot. With this dictum and with no clear objective for improvement, the task assigned to a robot is chosen assuming the robot will fail. If a production manager's company reputation might be damaged if a malfunctioning robot causes a lowering of production, the manager's first instinct is to place the robot where it easily can be removed and the task can be reconverted to manual operation. Usually, such a task is decidedly not a good application for a robot.

While top management may achieve its goal of experimenting with a particular manufacturer's robot in an industrial environment, the local plant has not used the robot to its best ability and would be hard pressed to show any clear-cut advantage gained by the conversion. An analysis of the robot's performance, quite understandably, might show a low efficiency or high cost/benefit ratio.

In this and other chapters, questions are included for classroom use or self-study. Some chapters include suggested projects.

QUESTIONS

1. What are the three main reasons that manufacturing jobs are converted to robotics?
2. Give four manufacturing tasks that, because of high technology, might be done better by a robot than by a human worker. Explain why.
3. Give three jobs that might be converted to robotics for the health and safety of workers.
4. Say that a robot could replace a human worker at one work station on an assembly line and could smooth out a problem bottleneck. However, the robot cost per year would be the same as the cost for the human worker at that station. Explain whether you would install the robot and why.
5. How does a decreased scrap rate at one work station affect the whole production operation?
6. Explain why a robot might provide an economic advantage over hard automation.
7. What is the "Expensive Toy Syndrome"?

CHAPTER 2

ANALYZING THE JOB

For you to determine if a manufacturing operation is a good candidate for robotics, the job should be broken down into basic processes. Then, each process can be analyzed with an eye toward robot use.

The following are the most common processes used for manufacturing durable goods:

BASIC MANUFACTURING PROCESSES

I. Component Manufacturing
    A. Material removal
        1. Saw
        2. Turn
        3. Mill
        4. Punch/notch
        5. Shear
        6. Etch
        7. Burn
        8. Grind
        9. Broach
        10. Polish/lap
    B. Material deposition
        1. Welding (when not used for assembly)
        2. Plating
        3. Hard coating
        4. Sedimentary deposition/paper making
    C. Material forming
        1. Press
        2. Mold/cast

    3. Draw
    4. Extrude
    5. Forge
  D. Compositional change
    1. Heat change
      a. Bake
      b. Heat treatment
      c. Thermal deburr
      d. Temper
      e. Anneal
    2. Chemical change
      a. Carburization
    3. Pressure change
      a. Isostatic pressing
    4. Radiation change
      a. Ultraviolet ink curing

II. Assembly
  A. Fastening
    1. Screw
    2. Bolt
    3. Rivet
    4. Dowel
    5. Clamp
  B. Welding
    1. Spot
    2. Arc
      a. MIG
      b. TIG
      c. Stick
    3. Flame
    4. Solder/braze
    5. Friction
  C. Gluing
    1. Anerobic (cure in absence of air)
    2. Solvent evaporation
  D. Press fitting
    1. Thermal interference
    2. Miter
  E. Entanglement
    1. Weaving
    2. Strapping
    3. Sewing

III. Coating/Treating
    A. Spray application
        1. Spray painting
        2. Flame hard coating
        3. Foam covering
    B. Dip application
        1. Pickling
        2. Plastic coating
        3. Dying
    C. Transfer application
        1. Brush painting
        2. Printing
    D. Controlled corosion
        1. Metal oxide covering
        2. Tanning
    E. Film coating
        1. Veneer
        2. Shrink wrap
        3. Bag stuffing

IV. Material Manufacturing
    A. Ore purification
        1. Iron smelting
        2. Aluminum reduction
        3. Metallurigical enhancement
    B. Natural gathering
        1. Logging/plant picking
        2. Stone quarrying
        3. Well pumping
        4. Animal by-products gathering
    C. Chemical formulation
        1. Plastic production
        2. Reactant solution
        3. Chemical mixing
    D. Forced separation
        1. Centrifuge
        2. Petroleum cracking
        3. Sifting
        4. Brushing

Table 2.1 shows the basic processes used in manufacturing two products—an automobile wheel and a shoe.

Table 2.1. Examples of Multiprocess Production

Step	Manufacturing Process
Automobile Wheel	
1. Turn ore into iron	IV A1
2. Iron to steel	IV A3
3. Forge rough wheel	I C5
4. Turn to size	I A2
5. Assemble rim	II B5
6. Notch mounting holes	I A4
7. Size holes	I A9
8. Antirust coat	III A1
Shoe	
1. Hide removal	IV B4
2. Hide tanning	III D2
3. Leather dying	III B3
4. Oil well pumping (heel)	IV B3
5. Oil cracking	IV D2
6. Plastic heel compound	IV C1
7. Punch heel	I A4
8. Punch upper	I A4
9. Cotton picking (thread)	IV B1
10. Cotton combing	IV D4
11. Thread spinning	II E1
12. Sew upper to sole	II E3
13. Glue heel	II C2

NEW OR OLD WORK?

A process being considered for robotics falls into one of two categories. Either it is new work—still in the design stage—or it is old work being done some other way. In determining if *new* work is appropriate for a robot, we must follow three steps. First, we must establish the basic functional unit under consideration, that is, one specific and limited job that the robot is to

accomplish. Second, we must ascertain the feasibility of a r
this task. Is the robot able to do the job? And third, we must just...
of the robot in terms of technology, safety or economy. Can the job be done
better by a robot than by other means?

FUNCTIONAL UNIT

In the modern assembly-line method of manufacturing, items being
produced move along the line and have work done on them at various
stations. At a given work area, one particular task in the manufacturing
sequence is performed. The operations occurring before may have readied
the part, and the task at hand may be necessary to prepare the part for some
later task. Still, at each station, a task is performed without regard for what
comes before or after. This setup is ideal for a robot. The task is limited,
the needed space is limited, and the work is repetitious.

In considering the use of a robot, we must also determine how the part
will be received by one station and transmitted to another. In a manual
operation, an operator can be instructed to pick up a part, perform his
assigned task and correctly place it on the moving conveyer. In a robotic
operation, the robot must be able to do the same thing.

ABLE TO DO THE JOB

After we have zeroed in on one task, we must determine whether there is
a robot that can do the job. Talks with a few robot companies should give
us the answer rather quickly.

After one or more robots have been selected for consideration, we can
begin analyzing robot feasibility on the basis of technology, safety or
economy.

If the job being considered for robotics is *old* work, we follow somewhat
the same analytical steps. The basic functional unit must be established.
However, we need not keep to the existing work breakdown. For example,
it may take one person to weld two components together and a second
person further down the line to clean up the weld with a grinding wheel.
Yet one robot may be able to both weld and deburr the part at the same
station.

Once the basic work unit has been established for the proposed robot, we
must determine robot feasibility. How far must the robot move in
performing its task? How much weight must it pick up? What types of
motions must it generate? A time-and-motion study could be facilitated
because a human worker is doing the task.

Again, we must analyze the proposed conversion on the basis of technology, safety or economy. This should be easier than with new work, as records should be available on work flow, productivity and expense of the existing production method.

QUESTIONS

1. List three basic processes involved in component manufacturing.
2. List four methods of assembly.
3. List three methods of coating or treating.
4. Give three steps necessary to determine if *new* work is appropriate for a robot.
5. True or false? A robot is better on a job that is creative and varied than on a job that is limited and repetitious.
6. In determining robot feasibility for *old* or *existing* work, is it necessary to keep to existing task breakdowns? Why?

CHAPTER 3

PREPARING FOR CONVERSION

When a task is to be robotized, considerable planning, scheduling and budgeting must be done.

PLANT SHUTDOWN

The decision must be made as to when the plant should be shut down for the conversion process. It is difficult, if not impossible, to integrate a robot with other machinery without the cessation of work. Usually, a manufacturing plant has some downtime during the year when a robot could be installed.

SCHEDULING ACTIVITIES

Scheduling must be done so the right people are ready at the right time. First come the experts who will prepare the floor or physically install the mechanical portion of the robot. Next come those who will hook up the power and the control cabinet. Those who will program the robot and any other associated equipment come later. Not only must the manufacturer deliver the basic robot on time but also any added equipment.

TRIAL RUN

Usually, when a robot manufacturer accepts the task of integrating a portion of the robotic system, a trial run of equipment is held at some facility outside the manufacturing plant. This is to demonstrate the abilities of the robot and its peripheral equipment prior to the stopping of the manufacturing process. If the robotic system does not function properly, the customer could decide against using the system or wait to try it on site when a factory shutdown period was available. What is to be avoided is creating a downtime period to install robot equipment, thereby preventing the manual operation from being performed, only to discover when the line begins anew that the process has been ill-conceived or some construction element improperly performed. Productive time must then be used to convert the process back to a manual one. Regardless of how fast this can be accomplished, there is a loss of manufacturing time and hence profit for the company.

ESTIMATING INSTALLATION TIME

When a robot conversion is planned, an estimate must be made on how long the process will take. If the robot conversion is estimated at 40 hours of effort and if the plant is to be shut down for only 60 hours of normal production, some overtime might be scheduled in case a time miscalculation has been made. If this is the first robot installation at a particular site and no experience is available to guide the time estimate, users who have installed a similar robot might be consulted.

Robot installation time can vary greatly with the manufacturer and the task to be given to the robot. In most cases, robot manufacturers will supply estimates regarding the time needed to install the basic robot and to equip it for a certain task. These estimates may include such steps as preparing a level, epoxy-coated surface on which to mount the robot and installing specialized hydraulic lines from one area of the work place to another.

Where estimates from the robot supplier are not available, the customer might refer back to its nonrobotic equipment installations. The robot, including control cabinet, should take about the same time to move into position on the shop floor as a piece of nonrobotic equipment of approximately the same size and shape. Once the robot has been firmly mounted on some type of base, other rules of thumb can be used to estimate the rest of the installation time. We might estimate that it will take longer to establish the input and output hookups that are a part of the robot than to install an equal number of electrical connections on a nonrobotic machine. However, installing the mechanical portion of the robot could be faster than connecting the mechanical portion of a similar yet nonrobotic machine tool. The robot can have its program altered quickly and its positions of reach changed minutely. Therefore, it is usually unnecessary to exercise extreme precision in the location of the robot in its work environment.

Another time estimate guideline that could prove helpful, in the absence of previous experience, is that a robot generally will take longer to program for a given task than would some machine such as a CNC mill. Because robots are so versatile and because the nature of their work requires a physical manipulation along the real time path, robots usually require a more generalized language and a greater amount of decision-making input by the original programmer.

At the time of robot installation, it is quite common to connect the robot to some type of coworking or overriding process controller. This is done oftentimes to augment the electrical capacity of the robot, giving it more memory and intelligence, or to expand the control of the robot to include other machines within the work station. When a process controller is to be part of the entire work cell, robot installation planning must also include time to program and debug the programs that both the robot and the process controller will use in their communication processes.

Additionally, robot peripherals must be tested as the robot is installed. Time must be allowed for the structuring of the peripherals as a part of the work station and for use with the robot. Peripherals are not only those pieces of equipment attached to and working directly with the robot, but also include all of the other smaller machines that aid the robot in its task. Among these are conveyers, racks, parts bins and pallets. Each of the supplying or removing devices with which the robot works—controlled by or passive to the robot—must be properly fitted and tested at the time of the robot installation.

A final time estimate must be made for the installation of safety equipment, such as a barrier around the robot. Although safety may have been planned for in the initial stages, it is sometimes not considered important enough to be a separate planning item during installation. In some cases, robotic equipment is barely proven before it is put into use. And ofttimes, planned safety equipment not installed during the robot installation is never installed at all.

ESTIMATING COSTS

The planning process includes budgeting for the installation and operation costs. When no prior experience is available, there are several criteria that can be used to estimate costs. We can use the actual installation costs of other nonrobotic machinery of a similar size and complexity. For instance, a large robot may require approximately the same expense to install as a fair-sized CNC mill. The budget is affected by the connections that must be established, how the robot is moved to the correct location on the floor, and the amount of programming and testing necessary.

Operating costs of the robot such as power consumption, maintenance and an on-duty operator are also expenses that have counterparts in nonrobotic machinery. The hydraulically operated robot consumes power at a rate based on its hydraulic power supply. If a similar power supply is used in the plant and information about it is available, we can use this as a guide for calculating the estimated energy operating cost for the robot.

Robot maintenance and operator costs, however, are usually quite different from those of comparable nonrobotic machinery. Because of the highly computerized nature of the robot and the quality built into a robot as a maintenance-reduction factor, there is little robot time spent undergoing repair in comparison to many corresponding pieces of equipment.

One decided advantage in the preparation of budgetary plans is the fact that most robot companies provide a period of free installation time and a warranty period during which the robot can be broken in. After these have expired, manufacturers usually provide a service contract guaranteeing service expense to be no greater than an acceptable amount.

QUESTIONS

1. Why can't most robots be installed in a manufacturing plant over the weekend and be ready for work Monday morning?
2. When a robot is purchased, what is the advantage of a trial run at a facility outside the manufacturing plant?
3. True or false? To avoid unnecessary confusion, the best way to install a robot is to shut down the plant until the robot is in place and working properly.
4. If a plant is installing its first robot, what sources might it look to to estimate installation time?
5. When is the best time to install robot safety equipment?
6. When a prospective robot user has no previous experience with robots, what guidelines can be used to estimate installation and energy costs?

CHAPTER 4

SELECTING A ROBOT

When the potential robot user wishes to select a robot to perform a particular job, there are different ways to go about it. Most often two techniques are used, or a combination of the two.

CHOOSING ON THEIR OWN

It is common for companies, particularly those with previous robotic experience, to collect information from robot manufacturers and use it to make a selection. Such a company must do the application engineering for the robot. How much weight must the robot move? What time cycle must it complete? Will a robot be able to handle assigned functions? Such are the questions the prospective robot user must answer for itself.

HELP FROM MANUFACTURERS

Another widespread method of robot selection is to delineate exactly what must be done by the robot and present this information to robot manufacturers, making a request for a quote. This is a common technique used by fledgling robot users who lack the experience to do the application engineering and selecting by themselves. This method is also used by many larger companies, even though they may have a great many robots in use. They choose not to use their own personnel to go through the rigorous process of selecting the best robot for a particular job, engineering the equipment that must work with the robot and handling all of the other details. Not only might the cost be prohibitive, they might not have the personnel ready, willing and able to do the job. It is generally more efficient for those employees acquainted with robot-type machinery and peripheral equipment to work directly with the robot manufacturer.

Another reason a company might ask the help of robot manufacturers is the location of the proposed robotic facility. It is common practice for a manufacturer to spread plants across the country or around the world. And while the company may have at one location personnel with enough experience to handle the robot engineering and selection process, at any given location there may be no one with the proper qualifications.

GATHERING INFORMATION

Once a potential user has decided to gather information from robot manufacturers, it should do so as soon as possible. How much will it cost to have a robot do a particular job? What kind of robot equipment is available at a cost range considered reasonable? Questions such as these generally can be answered quickly and with minimal effort by manufacturers. Case studies on different areas of robot use can be provided to the potential customer. Manufacturers also provide brochures giving such information as rate-handling speed and accuracy. However, manufacturers often are reluctant to quote the price for a standard piece of robot equipment because they are aware the cost of a system can vary substantially from this base cost. They also might be reluctant to divulge a base price without first gaining some insight into the potential customer. Discounts are quite common for large robot purchases or when there appears to be the potential for multiple purchases in the future.

GETTING A QUOTE

When a potential robot user makes a request for quote, it must delineate the work to be accomplished. What part is to be worked on? How much space will the robot be allowed? What is the time cycle in which the robot must function? What is the ceiling for the expenditure? The potential user also may specify that a robot have certain features not necessary for the job at hand. For example, a company may have a rather simple task it wishes to have robotized. Company personnel may know that this particular job eventually will become automated by some other equipment, so they want a robot which easily can be fitted to a different job in the future.

THIRD PARTIES

There are many more companies involved in the robot business as third parties and systems builders than there are robot manufacturers. For the most part, these are companies that have tooled for the large use of robots, have designed and fabricated special machinery, or have taken contracts for installing machines made by other companies. These are usually general contractors who release portions of the overall work to subcontractors. A real service is rendered by these third parties to fledgling robot users because they usually have the experience to help make proper robot selection and to bring a robot into usefulness quickly, possibly at a cost lower than if the user had dealt directly with the manufacturer.

COMBINING SELECTION TECHNIQUES

If other robots have been purchased by a company or if the company is familiar with making large capital investments, a combination of selection processes usually is used to purchase a robot. General information is gathered from a wide variety of manufacturers of robots and peripheral equipment. Knowing what is available on the market makes the selection process more reliable. As certain robots are seen to have the abilities and structures necessary for the job at hand, a limitation process begins. A few vendors are selected, and a request for quote is made to each. A request for quote will outline exactly the job to be done, but may not ask the vendor to quote all the equipment needed. For instance, the quote request may explain that the proposed robot must work with a spiral conveyer but may not ask for a price for this conveyer.

It is common for items such as robot grippers or sensing mechanisms to be supplied by the robot manufacturer because of their uniqueness to a particular robot. A gripper designed for one robot may not fit or work with a different robot. It is generally not wise to have an item closely connected with a particular robot quoted by a third party. However, there are companies emerging that make standard lines of robot peripherals, specifically grippers and vision-sensing equipment, directly adaptable to many of the common forms of robots. Often, a base product is available that has, as an option, mounting flanges and power connections adaptable to a wide variety of modern robots.

The request for quote is generally as specific as the customer can make it, every detail being delineated to prevent possible misunderstanding. When quotes have been received and scrutinized by the customer, the selection can be made.

NO-QUOTES

At times, even though a large number of requests for quotes are sent to robot manufacturers, only a few quotes will be returned. The robot manufacturer may wish to make a no-quote return. The reason may be a technical inability to solve the particular problem or a business inability to supply the equipment within the constraints specified by the potential user. Or there might be some economic consideration such as a refusal to deal with a customer with a bad payment record.

Whatever the reason for a no-quote, the robot manufacturer is taking some risk. One does not need to have an invitation to a party refused very often before he stops sending invitations to the refuser. So given the opportunity, a robot manufacturer usually will answer a request for quote.

GENERAL CONTRACTOR

When the combination technique of robot selection is used, the customer might act as general contractor in installing the robot. Many different companies may supply equipment, all to be brought together at the customer's location into a single operating system. However, each vendor is responsible only for its own equipment. Sometimes when a malfunction occurs, several different companies will send service people to correct a problem. It may be found that the problem does not lie with the conveyer. It does not lie with the robot. And it does not lie with the gripping equipment. Each subsupplier may be able to demonstrate to the satisfaction of the customer that its equipment is not at fault. However, this is of little consolation to the customer with the malfunctioning system.

For this reason, a robot user may wish to specify one supplier as a prime contractor, establishing this supplier as responsible for the entire system. This system responsibility can be the most expensive portion of a robot installation.

QUESTIONS

1. Give two reasons why a large company, already using robots at some locations across the country, might enlist the help of robot suppliers in selecting a robot for a new facility.
2. True or false? Questions on the type of robot equipment it has available in a particular cost range generally can be answered by a robot manufacturer with a minimum of effort.
3. During an initial phone call from a potential user, why might a

robot manufacturer be reluctant to quote a firm price for a particular robot?

4. Why might a potential robot user specify that a robot have certain features not necessary to the job at hand?

5. What are some of the facts the potential user should supply a robot manufacturer in a request for quote?

6. What is a *third party* to a robot sale/purchase, and what service does it provide to the buyer?

7. In a request for quote, why might a potential user not list all equipment necessary for the robot installation—for example, a spiral conveyer?

8. True or false? A user must always order robot grippers from the same company that supplies its robot.

9. Give two reasons for a *no-quote* by a manufacturer.

10. When several suppliers are involved in a robot installation, what is the advantage to the user of specifying one supplier as prime contractor?

CHAPTER 5

SELECTING A ROBOT COMPANY

Let's assume you have a task you want converted to robotics. You've analyzed the job, done some planning and estimating and have some possible robots in mind. Now, how do you select a robot company with which to do business?

For the answer to that question, we must digress a bit and look at the development of the robotics industry, what customers want in a robot supplier and why robot companies fail.

ENTERING THE ROBOT BUSINESS

Beginning about 1980, a large number of companies got into the robot business. There were several reasons for this. Perhaps the most significant factor was the potential for business. Reports by government agencies and independent study groups spoke of a potential world market for robots of billions of dollars per year. This market should present itself in relatively short order, offering those early investors a substantial profit. Even a small market share of the tremendously large robot business could give a successful company a return on investment equal to its wildest dreams.

Another attraction for new robot companies was the relative ease of entering the market. No one company had a corner on the market, nor what seemed to be a preferred solution. A company capable of designing other automated equipment could quickly establish its own robot design and announce plans to enter the market.

Still another attraction was the economy of scale used by robot manufacturers already in the marketplace in 1980. Early robot sales were rather meager in comparison to future estimates, and the equipment that was actually sold was quite expensive. The investment in research and development for a simple design could be recouped with sales of only a few robots. For the new company, this meant a great possibility for success and little risk on an untested product and market. Even the largest robot manufacturers were changing their products so often that the fledgling industry was in a constant state of flux. Each new design could be paid for on its own sales and did not require years of continuous production to return investor capital.

At that time, there was a fascination with high technology equipment. Although many businesses were unprofitable during those years, the companies involved in high technology products such as video games and computers generally were enormously profitable. Since robots were by their nature high tech equipment and since their very name inspired visions of a bold new future, new robot companies and large companies forming robot divisions found monies easily available.

Because of their high tech nature and potential for tremendous profit, many major companies in the industrial realm felt it necessary to procure or develop their own robot division. Nevertheless, such divisions owed part of their emergence to a "me too" element of corporate planning. This is not to deny that sound business judgment might have been used to develop such a department, but corporate fascination with a stable incorporating one of every major type of enterprise cannot be denied.

These and other factors led to the emergence during the early 1980s of approximately 50 new robot manufacturers in the United States alone. Many of these companies had powerful backing, technologically and financially. But as the research and development stage ended and the marketing stage began, many of the new ventures floundered.

WHO BUYS ROBOTS?

The robot industry, although supplied by a great number of vendors, serves comparatively few industries and companies.

The market for robots is concentrated in companies that have several factors in common: They pay relatively high wages; they're labor intensive and use repetitious manufacturing sequences; they can afford the capital investment for robot equipment; and robotic production of their products is technologically feasible.

High Wages

Industries paying comparatively high wages are prime candidates for the use of robots. Robots are a replacement of human labor and, where the production cost by labor is sufficiently higher than the cost for the same production by robots, robot use is economically feasible.

In considering wages, both benefits and insurance/safety requirements should be taken into account. An industry paying large benefits, even though wages may not be high, is still a prime candidate for robots. Likewise, an industry that has unhealthy or dangerous work environments should consider not only the primary benefit of preventing human illness and injury, but also the economic benefits of robots.

Such high wages can be found in the automotive industry, in heavy equipment manufacturing and in aircraft manufacturing. All of these industries make use of large numbers of robots. By comparison, the textile industry, meeting other requirements for an industry potentially open to the use of robots, does not pay such high wages and uses few robots.

Labor Intensive

Only companies using human labor as a major component of their manufacturing are prime candidates for robots. If most work is accomplished by complicated machinery or if most of the labor used is from outside the company—such as in the case of a subcontractor network—robot use is generally not feasible. For this reason, companies that show up in market reports as having large sales volumes are not necessarily large robot users.

The petroleum industry, which produces the same dollar volume as the automotive industry, uses comparatively few people. A handful of people can operate petrochemical equipment and produce millions of dollars' worth of product. The automotive industry for the same dollar-volume output, would require thousands of human beings or robots.

Repetitious Production

For an industry to use robots, it should have highly repetitious operations. Although robots are capable of being easily programmed, it is not economically feasible to program them continually. The robot jobs must be similar enough, day after day, so the programmer need not always be present. Some industries make products with anomalies precluding the use of robots. In some cases, very specialized machinery can be used, but changes in how the machinery functions are necessary on a continual basis.

Consider the building products industry. Although a particular type of wood or even a particular size of tree may be used in large volume at a sawmill, there are minor differences in wood itself. There are holes and knots, splits in the wood, and changes in the grain pattern that must be examined by human beings and taken into account for efficient production.

On the other hand, the heavy equipment manufacturing industry makes products that are the same at least scores of times before they are changed. Predictably, the building products industry does not use many robots, and the heavy equipment manufacturing industry does.

Manufactured items that have a short run, must be changed rapidly, and must be made using the same equipment are also not good candidates for robotic production, for example, women's ready-made garments.

The nature of the women's clothing industry is that styles change constantly. With high-fashion items for exclusive shops, quantities are kept relatively low. Because of the continual reprogramming that would be necessary for these short runs, robotic production would not be economically feasible.

Robots could be used to mass-produce standard items such as a basic blouse or skirt. But, since the garment industry does not meet the criterion of "high wages," production generally would be less costly by labor and nonrobotic machinery.

Capital Investment

For an industry to make use of large numbers of robots, there must be individual companies willing and able to make the necessary capital investments. Although there are many jobs in *food handling* and *processing* that could be handled by robots, the use of robots in *food production* is less feasible. A major reason for this difference within a single industry is the size of the individual operating units.

Food is generally produced by relatively small family-run units. An individual farmer might spend $200,000 on a combine to help harvest his crops but would not be willing to spend $100,000 on a robot simply to replace human labor. The efforts of the farmer and his family members would be considered sufficiently low in cost so that a robot could not be justified.

On the other hand, a robot capable of sorting eggs by size and without damage is an economic possibility for a large distribution company that sorts large quantities of eggs. What's more, the capital available to such a large company is generally greater than that available to an individual farmer.

Technological Feasibility

There are some industries not technologically suited for robotics. An example is the building industry. It meets the criteria of high wages, labor intensive, repetitious manufacturing sequences, and the ability to afford capital investment. However, robots are not yet capable of scaling heights by ladder, holding boards in position ready to be nailed, or laying brick that will conform to the lines of an irregular building.

WHAT CUSTOMERS WANT

Practicality, Long Life, Serviceability

Most robot customers insist on a robot designed to do the job, to last a reasonably long time and to be serviceable throughout its useful life. The reason is quite simple. High-volume manufacturing is economically intense—there is a great monetary penalty whenever production is stopped. As an example, in a plant where an automobile rolls off the assembly line every minute, each minute of lost production can be equated to the profit loss on one automobile. Therefore, a reputation for producing reliable, long-lasting equipment is very important in the vendor selection process.

A robot division or subsidiary of a company with a reputation for quality and longevity has a good chance for success in the growing robot market. Upon entering the robot market, reputable companies producing items such as machine tools seem to be able to transmit this reputation to their robot product.

Much like any other consumer product, robots come with warranties that they will be repaired during a certain period at little or no cost to the user. Beyond this period, of course, the customer is responsible for repairs. So, a robot company that can demonstrate that its products are easy to service can expect good customer reaction.

Robot companies sell serviceability by publishing data on the percentage of time their robots in the field remain operating. This is usually called *percentage of uptime*, and numbers above 98 percent are common.

Serviceability is often provided by creating a robot in replaceable modular sections. If an electric motor burns out, the entire motor assembly including feedback mechanism and power supply can be quickly replaced.

Corporate Longevity

Another thing customers want is assurance that the robot company will be around when their robot needs repairs. Large companies that are, without their robot divisions, profitable and growing are considered stable even if their individual robot products are not selling well. The marketplace feels such companies likely will remain in business, will be able to provide trained service personnel and will keep a quantity of spare robot parts.

Such is not the case with some robot vendors who merely have a licensing arrangement with some foreign company. The license may at some time expire or the local company may go out of business, forcing the robot user to buy spare parts from a foreign source or, if there is no supply, to have them custom made.

Update Equipment

Customers want their robot company to be able to make the research and development commitment to continually update robot equipment. Here, large companies have the edge. A small company, although selling sufficient product to be profitable, may not have the capital to compete on a world-class level. An analysis of independent companies in the United States shows that their products are generally at the low end of the technological scale. When a large company has a robot division, an economy of scale is reached in research and development allowing a greater return on research funds. For example, a good robot control mechanism might be used to control an NC machine or some type of process industry function like chemical plating or chemical refinement.

FACTORS LIMITING SUPPLIERS

Customer Training

Among the several factors limiting the actual number of robot manufacturers is the need of the customer to train a minimum number of people in the programming, repairing and installation engineering of robots. Even if the training period for a single robot is short—for example, a one-week intensive course for a robot programmer—the number of robots from different suppliers must be kept low so that the total number of weeks a robot programmer need be away from his job is low. Fifty different makes of robots would equal 50 weeks of training time. Therefore, robot users generally limit their suppliers, perhaps to three or four from different points of the robot spectrum.

Spare Parts

Another factor limiting the makes of robots is the spare parts necessary to keep robots continuously working. Each robot manufacturer has its own subvendors for equipment like motors, feedback modules and fasteners. For the most part, there is no commonality for this equipment. If a user limits its robot makes to three or four, it can keep its stockpile of spare parts to a manageable size.

Further, a company large enough to fabricate its own parts can create components that are interchangeable on a given robot. If a five-axis robot has a motor for each axis and these five motors are interchangeable, the customer need stockpile only one or two replacement motors.

Breaking-in Expense

Engineering experience during the breaking-in period can be very costly. The first time a particular robot is used in a plant, the customer's engineers learn how to use that robot effectively. Should miscalculations have been made in the initial design, or should misinformation have been supplied by the manufacturer, very costly retrofitting of components might be necessary for proper operation. Once this first installation has been completed, however, the engineering staff should be sufficiently trained so that the next installation will be much easier. If a company were to use a large number of different robots, its engineering staff would have to go through a learning period with each one.

Repeat Business

The robot industry, like many others, is built on lasting relationships between customer and vendor. In a first robot installation in a particular facility, the vendor may earn very low profit because of initial investments of travel, office rental, peculiarities in dealing with a particular customer, or specific product changes necessary to meet customer specifications. Most robot suppliers will endure the hardships of an initial installation, however, because one good installation will lead to another. When a particular vendor has proven itself, the customer will want to deal with that supplier again. This acts to reduce the number of robot suppliers in the business.

A good robot supplier generally will go all out to keep the customer happy. When there is the potential for new orders, when there is always in progress an engineering study that may provide more business for the robot vendor, there is an accelerated commitment to keep the robots already in place at the customer facility in prime running condition. When there is a critical malfunction in existing equipment during negotiations for a new and potentially lucrative robot order, the vendor views it as catastrophic. This may eliminate the vendor from consideration for the new business.

When a large organization such as an automotive company forms a long-term relationship with a supplier of robots or virtually any other type of equipment, this state of critical negotiation time is extended to a continuum. Over the years, there may be no time when it is not critically important for the supplier to keep its previously installed equipment in prime running condition. Suppliers might speculate that this is done intentionally — that a constant state of crisis is maintained by the user to get the maximum service from the supplier.

Purchasing Expense

Still another factor limiting the number of robot vendors in the marketplace is the purchasing expense. Companies large enough to purchase robots generally have purchasing departments. The department facilitates low-cost acquisition of goods and attempts to keep to a minimum paperwork, meetings, and wrong shipments. In purchasing robots, like other materials, the actual purchasing process doesn't cost much more for quantities than for a single item. To purchase a $100,000 robot, a customer might spend $25,000 in wages, benefits and facilities use. The purchase of robots valued at 10 times that amount (which might be 11 robots, allowing for discounts) would not require 10 times the purchase process expense. The expense might only be doubled. This provides a tremendous economic incentive to deal with only a few companies rather than spread the business among a multitude of suppliers.

WHY ROBOT COMPANIES FAIL

Bad Timing

Bad timing in entering the marketplace has led to the failure of several robot companies. There is only a certain *window* through which a company may emerge into the market and be a strong factor. This same phenomenon can be seen in other industries—the automobile industry for instance. There is no major American automobile company in existence today that started later than 1930. Certainly, some suppliers in business prior to 1930 have also gone out of business. But to be a forceful contender in the modern automotive manufacturing market, a company needed to have found the marketing window previous to 1930 and to have established a place for itself in the minds and buying habits of car buyers.

At the beginning of the robot industry—the mid-1970s—a variety of robots were tried. New and unproven ideas were accepted. But as the market matured, the marketing window grew smaller and smaller. Large companies established themselves as strong forces in the marketplace. It became more difficult for newcomers to overcome the competition.

Insufficient Financial Backing

Companies entering the robot market with limited capital hoping to gain a quick return have not fared well. There were some companies formed on a shoestring and expanded by stock sales, loans, and reinvestment. However, the robot industry is not well suited for this growth technique. A company must invest in advertising and marketing before it can sell even a modest amount of product. It must invest in knowledgeable sales people to make daily contacts to sell the product, the company, and themselves.

To get even a first robot sale in the auto industry might require thousands of hours of meetings, demonstrations and paperwork. Once this sale is made, of course, each sale becomes easier, both for the supplier and buyer. Robot companies starting off with little backing sometimes misjudge the amount of legwork necessary to establish one of their robots in a major company. Even a delay of a few months can be disastrous to an under-funded company. Many small companies that did not fail have become divisions of large corporations.

Poor Design

Another reason for company failure is poor design. Many companies are formed by business people who have little understanding of robots. This need not be a problem as they can hire consultants. However, in the past, through poor communication, erroneous assumptions and other reasons, robots were designed that were ill-suited to industry. Some of these seemed to be very innovative and ahead of the rest of the industry. Generally, they were more anthropomorphic and more stylized than their mundane but more widely used counterparts. Yet, they tended to be impractical for industry. For example, they had certain design flaws such as an inability to work in demanding environments or a requirement of a long downtime for servicing. Companies that did not have a firm technical grasp of the end uses of robots sometimes found their product virtually unsaleable.

Inadequate Service

Robot companies have failed because of the lack of adequate service. It does not matter to customers whether they are buying a company's first, or one thousandth, robot; they want repair people ready when their robot malfunctions and halts production. A beginning company has to put together from scratch a staff of service people. And training of these people is sometimes a problem, as there are few units running in real industrial locations where they can gain experience.

Those companies that could quickly train and field a support organization found acceptance in the marketplace not necessarily because of a superior product, but because of an overall operating efficiency based on human and robot factors. Indeed, a robot that would malfunction more often but could be quickly repaired was usually more saleable than was a robot that malfunctioned less but remained inoperable for long periods of time.

Inadequately Trained Customers

Still another reason for robot company failure was the inability of customers to program and repair the product. For example, some robot companies offered a training course but, upon completion of the course, the customer's employees were unable to perform the specified procedures.

It is important that a customer's staff be thoroughly knowledgeable about its robot. When a customer buys a robot to perform a specific task, the robot is engineered by members of the supplier's staff who are intimately acquainted with the machine's peculiarities and advantages. As time goes on, however, and the robot must have its task changed or be reassigned to other tasks, the customer's plant personnel must make these changes. Without an adequately trained staff, mistakes are made. It then appears to the customer that the robot is improperly designed or difficult to use.

Improper Marketing

Some robot companies with a good product and a good distribution and service organization had problems in the marketplace because of improper marketing or local representation. As an aid to entering the market quickly, many companies decided to sell their robots through distributing agents. Usually these were agents who were acquainted with local customers and who sold automated equipment and other machinery. There were some drawbacks, however.

Sometimes a distributor was chosen who handled products in direct competition with the robot equipment. When a particular distributor has the option of selling one piece of equipment or another, he may favor the one best suited to a customer's needs. On the other hand, he may sell the one earning him the best commission. What's more, a salesman *sells* best what he *knows* best. A salesman might be well versed in automated equipment but not fully understand robots. Therefore, he would be less skilled in selling robots.

In some cases, a particular manufacturer's robot sells well in some areas of the country—or even in some segments of a particular market—and fares poorly in other areas. Often, this is the result of a few people improperly marketing the robot. A distributor calling on the east side of a town might lead to the development of a bad reputation for a robot. At the same time, a different distributor on the west side might help it develop a good reputation. Information about the proper and profitable use of a product travels slowly, but bad news travels fast.

This fact is amplified in large companies with manufacturing plants scattered throughout the country or around the world. At any one time, a large automotive company might have 20 different distributors selling to it, many representing the same robot companies. An overly aggressive marketing approach by one distributor might place the robot in a borderline application. Then, if malfunctions occur or if a low efficiency rating is achieved, that information is distributed throughout the organization.

Lags Behind in Technology

Another reason why a robot company might fail is its inability to lead or keep up with high technology. A large, robust manufacturing company is usually able to make a research and development commitment leading to product at the high end of the technological scale. Indeed, much of the equipment and knowledge necessary to make a technologically advanced robot is already in place.

Smaller companies entering the robot market and drawing on existing knowledge in the field usually find their products are one to two years behind those of the leading companies. With the state of the art expanding so rapidly, a perception by the marketplace that a particular robot is outdated is oftentimes sufficient to eliminate it from consideration by potential customers.

Some companies purposefully aim for the low technological end of the robot business. These companies explain their product as a low-cost alternative to a high-technology robot. Some large suppliers of robot equipment have successfully taken this approach. However, a company now entering the market and wishing to use this sales technique would be hard pressed to show any product advantage. If it is not high technology being sold, then a proven simple mechanical robot and control system from an existing manufacturer probably would be preferred by the customer. Further, a new company entering the market— because of initial capital investment—probably would have to price products higher than established companies that have their manufacturing operations in place.

THE IDEAL ROBOT COMPANY

Learning from past failures, we should select a robot company with a strong financial backing, often from a large, well-established parent company; a well-designed, practical product; good service, training and marketing operations; and an ability to lead in or keep up with high technology.

QUESTIONS

1. Give three reasons why many companies got into the robot business in the early '80s.
2. What is the "me too" element of corporate planning?
3. List three factors that companies using robots generally have in common.
4. List two industries that use a large number of robots.
5. Why isn't the textile industry a leading user of robots?
6. Why isn't the petroleum industry a leading user of robots?
7. Why isn't the building products industry a leading user of robots?
8. Why don't more farm families use robots?
9. Give an example of an industry not *technologically suited* for robots.
10. True or false? Robot reliability is important in automobile manufacturing because there is a great monetary penalty when production is stopped.
11. Explain the meaning of *percent of uptime*.
12. What are two risks in purchasing a bargain robot from a company going out of business?
13. What is the advantage to the robot user of its robot supplier investing in research and development?
14. List five factors that tend to limit the number of companies in the robot business.
15. Give seven common reasons why robot companies have failed.
16. List four attributes we should look for in a company from which we purchase a robot.

CHAPTER 6

MAINTAINING THE ROBOT

The maintenance of a robot really consists of separate functions often lumped together under the category of service. These involve such things as routinely boiling and cleaning robot machinery, repairing the robot when it is broken, and overhauling components when they are worn out. These different functions are performed by different specialists under entirely different conditions and with different economic incentives.

ROUTINE MAINTENANCE

Routine robot maintenance often involves lubrication only. On a semi-regular basis, oil is added and changed— transmission oil and grease for an electric robot and hydraulic oil for a hydraulic robot. In a large manufacturing facility, this work is usually done by persons called oilers. These people keep records on each piece of machinery and apply oil and grease according to a predetermined schedule.

While routine maintenance procedures are designed by robot manufacturers to facilitate proper lubrication of mechanical components, they are often inadequate. Varying conditions at a particular robot location, such as temperature, amount of contamination in the air and specific use of the robot, preclude any realistic broad-based determination of the lubrication requirements for the piece of machinery at hand.

It is not feasible, however, in a large, busy and noisy manufacturing facility for an oiler regularly to look at each piece of machinery to determine its need for lubrication at that moment. Many plants are so noisy that a component screeching due to the dry fit of some components cannot be heard. A component that has slowed or developed a stuttering action because of the stickiness of its parts likely will go unnoticed if it still performs its assigned task.

Some manufacturers have one or two oilers employed full time to lubricate machinery on a continual basis. When, as is common, they are constantly interrupted to perform other functions (such as buying new batteries for the cafeteria radio before the baseball game), the maintenance is not completed as scheduled. Either some equipment must be omitted or all of the equipment must be pushed to an extended schedule.

What's more, there is little job satisfaction in oiling when it is often impossible to tell if the job is being done adequately. Quite unlike the old-time train engineer, oil can in hand, who attended to the wheels of his train while it was in a station, the modern-day oiler pushes about a cart containing pressurized vessels of oil that he inserts hypodermically into the machinery. (Some oilers displace large amounts of oil into small cavities—and subsequently onto the floor—and yet completely ignore a needed application, resulting in machinery loss.)

Oiling is not what most people think of when they analyze the performance of a robot and its need to function economically on a continuous basis, yet inadequate lubrication results in the poor use of expensive machinery. A component designed by astute engineering minds to have a continuous service life of 10 years might fail after two, should it not receive the proper lubrication.

A current school of thought in engineering and design is that all designs must take into account a complete failure to lubricate by the user. A component such as a gear, which inherently has a scraping edge of metal, is designed to contain permanent lubrication to guard against customer neglect.

Another often-overlooked function of routine maintenance is the cleaning of electronic components when they are exposed to a contaminated environment. Many robots have electrical control units in cabinets that have access covers sealed with rubber strips. These control units, because of the hot factory environment, are commonly cooled by fans or air conditioning units. The air flow causes certain electromagnetic attractions to occur inside the cabinet. Contaminated factory air containing minute particles of metal dust is attracted to the electrical components of the control unit.

Many components such as those built with a CMOS structure are very sensitive to such contamination. When even a single human touch can destroy the electrical functioning of a CMOS component, one can imagine the damage caused by a one-eighth-inch layer of magnetized metal dust conducting electricity between all of the available pins on the back of a circuit board. Some robots have completely destroyed themselves by electrical fire—the only plausible cause being a buildup of conducting-type metalized dust within the cabinet.

A useful regular maintenance procedure is to open the access doors into the robot and use a vacuum cleaner to remove dust particles from the cabinet. Sometimes, following manufacturers' directions, a maintenance employee can use Freon or another such solvent to remove stubborn stains from circuit boards and other robotic components. Many manufacturers are reluctant, in their service literature, to indicate the necessity of cleaning a control cabinet interior. However, where this is a common practice, a greater service life is achieved for the equipment.

A robot designed with double-sealed components or with a cooling and ventilation unit (allowing those components producing a great deal of heat, and yet less subject to contamination, to be isolated from more sensitive components) has a greater chance of survival in a dirty and hostile factory environment.

ROBOT REPAIR

There are basically three groups of people that perform robot service. There are those who are employed by the user company specifically for the repair of robotic equipment or who serve in some capacity, such as an engineer, that allows them to make repairs.

Secondly, there are those employed by the robot manufacturer. These might be designated robot repair persons or any personnel fielded by the robot manufacturer who are capable of rendering assistance in an emergency. The vendor's service manager, salesmen, sales manager, engineering staff personnel, research staff personnel (everyone except possibly the company president or controller) are apt to be pressed into service when a customer needs robot service.

A third group performing robot service is third-party suppliers. These are generally companies specializing in machinery repair or independent contractors who repair a wide variety of machinery and electronic devices. Only those robot companies with a sufficiently large number of robots in the field to warrant independent service companies learning their equipment can rely on third-party suppliers.

ORDER OF SERVICES RENDERED

The first people to attempt robot repairs are usually the customer's employees. In almost all cases, someone at the customer's facility is acquainted with at least the simplest robot repair procedures. This person might be an engineer or electrician. If qualified customer employees are not available, can't handle the problem, or would be overburdened by the number and complexity of repairs, the customer then looks elsewhere for help. Because of warranty agreements and in some cases contractual requirements, it is usually the robot vendor who is contacted first. Only in such situations as a history of bad performance by the robot vendor or a location and response-time advantage would the third party be contacted first.

SUBSTITUTE ROBOTS

There are times when even the three-layered approach to repair service is inadequate. Robots are, after all, extremely complex mechanical and electrical devices, and modern man has not exorcized all of the robotic poltergeists in American industry. A simple malfunction occurring spontaneously might take many days to repair, even though great effort is expended by customer and vendor service personnel.

In these cases, an increasingly common practice is to have ready a completely separate robot for use in a particular work station. Thus, no malfunction, regardless of how serious, need shut down the plant operation for very long.

In some cases, specific programs are developed by robot manufacturers so a previously written program might be loaded off one robot and into another, then updated with only a few simple operations. It is possible with these systems to take a program containing perhaps 100 positions in space and, by altering positions 1, 50 and 100, for example, allow the machine to completely rewrite the program so it is adaptable to the specific differences on the second robot.

Some robots allow the exchange of one control system for another and the operation of a previously assigned program without alteration. In more advanced robots, a sensing mechanism can be used that automatically identifies for the new robot the changes in mechanical structure or feedback control necessary before it can begin operation.

In most cases, an extra robot could be available in anticipation of a major malfunction. There could be a regular exchange of robots, allowing each one in turn to be routinely serviced and gauged. This is not a common practice, however, because of the cost of the extra robot in comparison to the cost of stockpiling modular spare parts and making repairs on the robots in service.

DIVISION OF DUTIES

Like service on other plant machinery, robot service is usually divided according to trade boundaries. The repair of electrical machinery is generally performed by personnel different from those who repair mechanical machinery. Also, because of union designations in some organizations, different levels of programming functions are performed by different personnel. There is little commonality among organizations as to the divisions of labor and indeed, in each manufacturing facility, unique anomalies develop to the consternation of outside vendors.

In some plants, it is the job of electricians to repair virtually all of a modern electrical robot, since there are few mechanical components separated from electrical motors. In other facilities, since the robot is viewed as a mechanical device that is simply powered by electrical motors, it is a machine repairman's function to make all repairs outside of the electrical cabinet.

Then, there is *incidental* work—work that is only incidental to the performance of a major task. For example, if a steering component requiring a half-hour repair time by a machine repairman is to be disassembled and if there is an electrical connection that must first be disengaged, this simple electrical task might be considered incidental work for the machine repairman. In other words, it might not be necessary for two people to make the repairs.

Also, in the assignment of service tasks, distinctions are made based on the size of components, their weight, their speed and their function. In a large manufacturing facility, motors in the two-horsepower-and-above range might be serviced by personnel different from those working on smaller motors. In a small facility, such distinctions are not made. However, since robots are used primarily by the large manufacturers of automobiles, heavy equipment and aircraft, and these companies are the very footholds of American labor unions, debates as to the proper demarcation of trade work are common.

Electricians, likewise, do not all fit into the same group. Some are specifically trained and have trade classifications allowing them to service computer-type components. Others may service installations with very high power, and potentially lethal, electrical voltages. In many cases, the power supply is taken into the robot at 440 volts AC, a voltage potential considered dangerous to anyone not trained specifically in high-power electrical functioning.

To further complicate the division of duties, robot service might also involve different trades to handle the pneumatic or hydraulic power connections and pipe fittings, the mechanical bridging networks and heavy-lift equipment such as chains and hooks, and programming functions to interface different pieces of equipment.

HORROR STORY

There is, in the robot industry, a story told at great length by virtually every serviceman. The plant location, personnel and the robot company change, but the story, usually told in local taverns, goes something like this:

"We had a simple repair job involving the removal of a connecting pin from a motor element. An electrician had to disconnect the high power. A machine repairman removed the electric motor. A different electrician disconnected the low-voltage power at the motor. Still another electrician turned off the computer functioning of the robot. A pipefitter removed a connection involving a certain type of thread. And an engineer supervised the overall operation.

"The repair took several hours, allowing for coffee breaks, lunch breaks, miscellaneous misunderstandings, a failure to share tools, and at least one argument about work demarcation. I could have done the whole job myself in a few minutes."

PAY SCALES

The personnel performing repair functions both inside and outside of the user company are often highly paid. While it is difficult for a person without a management position or advanced degree to achieve the earnings potential of an American automobile worker, some repair personnel, both for robotic equipment and other expensive or complicated machinery, are among those who do. It is not uncommon for a repair person in the automobile industry, with almost automatic overtime, to earn some $45,000 per year. It is uncommon, however, for a repair person outside of a large automotive or heavy-equipment manufacturing company to earn this much.

The reason for the differential is this. The repair person operating within the large manufacturing company is, for the most part, unionized and paid at an hourly scale that allows for overtime. On the other hand, since robot companies are relatively new entities, their repair people are relatively new to the companies and new to the industry. The service people have had little chance to reach a high earnings level by working many years with the same company or by, over time, selling their services to increasingly higher bidders. What's more, these service people generally don't earn as much in overtime. They might be asked to work long hours when necessary, then given relief time when their services are not needed.

The third-party robot repair people are usually paid on a scale commensurate with their abilities and in line with wages of robot service people employed by users.

ROBOT OVERHAUL

When a robot has been in service for a considerable length of time, substantial repairs, especially of mechanical components, might be required. The robot might have to be removed from the assembly line and even from the factory itself. A robot overhaul, which might include the disassembly of bearing components or swivel joints, can take days and usually can be done most easily at the original manufacturing facility. Many specialized tools and fixtures are necessary to retrofit bearing components properly or to remachine components to correct tolerances and specifications. When it is impractical to ship a robot back to its manufacturing facility (such as when a robot was made in a foreign country), user facilities can be outfitted with the tools to perform the overhaul. In some cases, it is considered faster and cheaper to scrap large portions of a robot and retrofit complete new sections rather than repair failing components.

This conforms to the American philosophy of using machinery as an expendable item, as opposed to the European (and Oriental) philosophy of good routine care of equipment so that it may last a long time.

REPAIR OR REPLACE?

Large-scale manufacturing—once it developed in the United States—brought with it the idea that machinery was functional only so long as it was easily serviceable. And, since machines would be replaced anyway as improvements were made, it was not considered prudent to keep antiquated machinery running at all costs. This attitude is prevalent regarding not only factory machinery but other products.

A five- or six-year-old automobile, showing signs of wear, can be purchased for a small fraction of its original cost. Ten-year-old automobiles are relegated to the lowest economic class and teen-agers.

In some other places of the world—Europe, Japan, South America—an automobile that is in running condition and a mere 10 years old still possesses considerable value. It is not uncommon for a European to have an automobile engine completely overhauled three or four times, and, as any taxi rider in a South American city can attest, a mere 200,000 miles on an automobile is no reason for a disparaging attitude on the part of its owner.

PRODUCT OBSOLESCENCE

The American idea of product obsolescence is prevalent among robot users, and with good cause. Robotic equipment designed in 1970 to have, with proper care, a useful life of 15 years, is of almost no value today on the used-machinery market. This is because of cumbersome control systems, energy inefficiency and nonmanipulative mechanical structures. The equipment sold in the 1980s surely will be replaced before 1990 by less expensive, more productive and easier-to-repair machinery.

In the 1970s, a robot component with a large-scale computer controller could easily cost, in preinflation dollars, $100,000. In 1980, a component that could outperform the original could be purchased for perhaps $35,000. It is expected by both robot vendors and users that equipment being purchased today will never wear out. Not that it is incapable of wear, rather it will not be used long enough to wear out.

HOW MANY SERVICE PEOPLE?

The amount of service required by robots in a facility varies greatly depending on their assigned tasks and manufacturers' care in designing and manufacturing. There are, however, certain rules of thumb to estimate the number of service people needed to keep them in operation.

As a general rule, one repair person might be able to keep running 25 or so robots. This is based on the percentage of uptime of most modern robots—between 98 and 100 percent. It also assumes that the 25 robots are in proximity to each other and at the actual work site of the service person involved. One service employee per 25 robots means, if there are three shifts, as many as five people might be employed specifically to repair robots so each shift is covered and there is allowance for absence and vacation time. These might be four regularly scheduled repair people plus one supervisor who would aid in repairs. This would allow for emergencies when two people might be needed to repair a robot or when several robots might malfunction at the same time.

When the same number of robots are to be kept working by vendor servicemen, generally more people are needed. Travel time and warranty terms must be considered. If a manufacturer warrants that, after a request for services is made, no more than three hours will elapse before a service-man arrives, obviously service people must be stationed throughout the area of robot use. If robots are located close to each other, such as when several are in a single plant, only a few people might be necessary to service the area adequately. When the same number of robots are in far-reaching areas, a greater number of service people would be required.

For this reason, many emerging robot companies are at a decided disadvantage in the area of service. A robot company that has sold 25 robots scattered throughout the United States would perhaps need 10 full-time service people to assure that a three-hour service response time could be met. The same number of personnel might service perhaps 200 robots if the robots were more densely located, perhaps several at one site. The older robot companies and those that have dealt in a substantial number of products are more likely to have sold multiple robots to many companies. Indeed, some manufacturing sites employ between 50 and 100 robots in a continuous operation to perform arc welding or spot welding and have a resident vendor service person at the site.

Fortunately for the robot vendor, concentrations of high-technology manufacturing have caused the development of pockets of robot use. Figure 6.1 shows a map of the United States and the regions that tradition-ally show up in analyses of service needs. A particular company might have concentrations of robots in other places, but most service is needed in the areas shown.

Great need
Moderate need
Low need

Figure 6.1.   Service needs.

As an alternative method of service, some users join forces, share funding and develop a support group whose members share service people, spare parts and tools. Similar cooperative strategies have worked in other industries, such as the textile and tobacco equipment users in southeastern United States.

At times, local governments become involved in the servicing of robots or other equipment. Community colleges train robot service personnel and help set up operations wherein equipment might be repaired efficiently by a third party.

QUESTIONS

1. Give three factors that could affect the amount of lubrication needed by a robot in a particular location.
2. Give two reasons why routine lubrication of robots might not be performed by an oiler as scheduled.
3. What problem could result if, when there are metal particles in the air, the robot electrical control unit cabinet is not cleaned?
4. What are the three categories of people who repair robots?
5. Under what condition might a substitute robot be placed in a work station?
6. Explain *incidental work* as it relates to robot repair, and give an example.
7. True or false? When it comes to robot repair, there is great commonality in the industry as to the divisions of labor.
8. True or false? Often, it is faster and cheaper to retrofit new sections of a robot rather than repair failing components.
9. Give two reasons why robotic equipment sold some 14 years ago might have little value on the used machinery market today.
10. What is the percentage of uptime of most modern robots?
11. Give two factors that could influence the number of service people needed to keep 25 robots running properly.

CHAPTER 7

ROBOT TRAINING

Training in robotics is quite different from training needed for other manufacturing machinery. There is a newness in robotic machinery that precludes a large number of trained people. Robot machinery, although based upon technology found elsewhere in industry, has been developed only within the last few years and has been widely used only within the last three.

Aside from the newness of robotics, there is the sophistication of its technology. Robot technology is based on computer technology. However, most computers do not command motion in a device other than a hard-copy printer or magnetic storage device. Robot computers command movements of large weights and high velocities. Therefore, for safety reasons, personnel involved with robots need to be especially well trained.

Another reason why robot training is different—aside from the robot's newness and the sophistication of its equipment—is that the whole industry is relatively new. The new robot manufacturer has a training need not only for its engineering and production staff, but also for sales, marketing and other personnel.

Consider the sales department. While salesmen in the past might have sold equipment similar to robots, limited robotic knowledge might lead them to pass along erroneous information to customers. So, specific training in the abilities and functions of robotic equipment is necessary.

WHO SHOULD BE TRAINED?

For a robot operation to work smoothly and remain profitable, there are many different customer employee groups who should be trained. Among these are the engineers who will work with the robot machinery, the plant personnel who will operate the machinery, the plant supervision who will oversee the operation, other production personnel who will work in related operations, the personnel staff that must hire specialists and handle robot-related personnel problems, the purchasing staff that must buy the robotic equipment, and all personnel who might perceive the robot as a threat rather than an aid to production.

It is obvious that many of these groups will be trained because they will work directly with the robotic machinery. It is a somewhat novel idea, however, that groups such as personnel, purchasing, and even all company employees should get at least some training.

Because there has been so much written—both fiction and nonfiction—about robots, there are many people who are confused, alienated and in some cases afraid of robotic equipment. Thus, an effort by a robot user to educate all employees who might come in contact with the robot or who must make decisions about the robot can yield a high return.

A lack of training could cost the company considerable money. If a maintenance worker does not know how to maintain the robot properly, he might cause the robot to stand idle sometime in the future. Particularly in a union setting, if a worker misperceives the robot, he might create some disturbance, resulting in lost production time or damage to company property.

Training practices vary from company to company. A study of several robot installations indicates that, in some instances, little or no robot training was given to plant personnel. In other instances, virtually every person "with a need to know" was trained, and very aggressive programs were established to update workers' skills regularly. In some cases, the robot user has little choice about who will work with a particular piece of robot equipment. Constraints of seniority and trade designation will fix the selection. In other cases, however, a selection can be made based on employees' ability, desire and proven experience in robotics.

Because robots are relative newcomers to the manufacturing environment, many people usually will ask to work on them and show a great willingness to add to and upgrade their skills. This desire to work with new equipment and gain an important learning experience is not limited to any segment of the population. There is no proper age at which to learn about robots. In one instance, a 59-year-old factory electrician was assigned to learn electrical troubleshooting on a very sophisticated microprocessor-controlled robot. To many, this was an unwise assignment. The worker, however, was highly motivated to work with the new and exciting equipment. After a short training period, he became a successful robot repair person. He was able to learn the new microprocessor technology some 30 years after having served his electronic apprenticeship.

Likewise, neither sex has any advantage in learning about robots. Both men and women have been trained successfully in the programming and maintenance of robots. Robots do not care whether the person pushing the buttons is old or young, male or female.

WHAT KIND OF TRAINING?

Naturally, different categories of personnel need different amounts and kinds of training. Many robot companies offer a series of educational programs to their customers that includes these types of training:

1. Awareness
2. Purchasing
3. Production use
4. Mechanical and installation
5. Electrical and troubleshooting
6. Routine maintenance

Table 7.1 shows potential groups to be trained and the type of training recommended for each.

WHEN SHOULD TRAINING BEGIN?

Because of the long selection process and long lead time necessary to order robot equipment, it is usually unwise to begin robot training too early. If user personnel are trained several months before the robot arrives, by the time the robot is ready for use they may have forgotten much of what they learned and need retraining.

It is also a good practice to avoid delaying training too long but rather to capitalize on workers' curiosity and interest. A comprehensive training program—even one involving a large number of people—generally can be conducted over several weeks.

TRAINING COSTS

Robot training, when supplied by a vendor, could add up to as much as 10 percent of the entire robot installation cost. While training on a piece of robotic equipment costing $100,000 might cost the user $10,000, the savings generated by proper robot use can more than equal the training cost.

Table 7.1. Who Needs What Training

Employee Groups	Training				
	Extensive Technical	Simple Tech	Operation	Features/ Economies	Awareness
Process Engineers	X				
Tooling Engineers	X				
Material Handling Engineers		X			
Plant Engineers		X			
Engineering Staff Designing the Product		X			
Maintenance Staff		X			
Production Operation			X		
Production Supervision			X		
Personnel Staff				X	
Purchasing Staff				X	
General Plant Personnel Working Near Robot					X

QUESTIONS

1. True or false? For safety reasons, persons involved with robots should be especially well trained.
2. True or false? A successful salesman experienced in selling dedicated machinery will require no special training to sell robots.
3. Give five customer employee categories that might benefit from robot training.
4. Give two reasons why customer personnel department employees might benefit from robot training.
5. Why might a robot user have little choice about which employee will work with a new robot?
6. When a user is able to select the person to work with a robot, what are two attributes that might be sought in the chosen employee?
7. True or false? Both men and women have been trained sucessfully in the programming and maintenance of robots.
8. True or false? Employees are best able to grasp and retain robotic knowledge when they are between the ages of 22 and 35.

The best way to learn about the application of robots to practical work is to follow a robotic conversion through the feasibility study and the planning, selecting, engineering and installing processes. In the following pages, there are several examples of tasks converted to robot work. Although the conversion applications are based on real cases, details have been changed and eliminated to avoid giving preferential treatment to any manufacturer and to avoid giving confidential information about any actual robot user.

CHAPTER 8

LOADING AND UNLOADING A LATHE

The Standard Ball Joint Company made standard metal components used in steering and transmission mechanisms of heavy off-the-road construction equipment. In its plant, an operation 20 called for a steel bar to be turned, on a lathe, into the shape of a ball on the end of a shaft. Figure 8.1 shows the steel bar before it was machined; Figure 8.2, the same part afterward. Table 8.1 shows the manual procedure the operator followed using existing equipment.

Figure 8.1. A steel bar before machining.

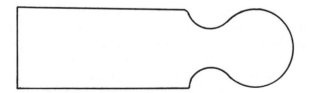

Figure 8.2. A finished ball joint.

Table 8.1. Manual Procedure

1. Get part from rack
2. Open chuck
3. Exchange parts
4. Close chuck
5. Start lathe
6. Put good parts in basket or
7. Put bad parts in small basket
8. Go to step 1

The part being machined was quite heavy—its weight before machining was 39 pounds. The time to load and unload the machine was 37 seconds. The machine cycle, during which the operator was idle, was 63 seconds. Each part made had a cost of 33¢. Engineers, noting the loading and idle times and in an attempt to lower the overall cost of machining this particular product, considered installing a robot.

ROBOT SELECTION

The task to be robotized was loading and unloading the lathe. Through other work in the plant, the engineers were familiar with several simple pick-and-place robots. They selected the robot shown in Figure 8.3. A sample robot program (Table 8.2) was devised to prove the technical feasibility of the conversion.

The price for the robotic equipment was $22,500. Based on their previous experience, the engineers decided that the equipment could be installed and programmed for an additional $5,000.

The robot purchase was justified when it was determined that should the robot operation remain viable for two years, the overall price per part would be 15.5¢, or a savings of over 17¢ per part. The calculations to compute the relative cost of each of the parts are shown in Table 8.3.

The particular robot selected was chosen on the basis of a successful working relationship between the manufacturer and the Standard Ball Joint Company and a price comparison of equipment available from several manufacturers. One option added to the robot was a base allowing the robot arm to have its lower position at the proper elevation to pick up the part directly from the cut-off saw. The equipment used with the robot is shown in Figures 8.4 – 8.10. Figure 8.11 shows a floorplan of the installation.

The input and output signals—from the robot to the surrounding equipment and from the other equipment to the robot—are shown in Table 8.4. At the time of installation, all these signals had been properly documented, and the robot took less than one day to install. The exact position of the robot on the shop floor was critical, since this position was not adjustable.

Table 8.2. Robot Program

1. Arm in
2. Grip open
3. Swing left
4. Down
5. Slide right
6. Arm out
7. Wait for signal "one"
8. Grip
9. Up
10. Arm in
11. Swing center
12. Down
13. Arm out
14. Slide left
15. Output signal "A"
16. Grip open
17. Wait for signal "two"
18. Slide right
19. Arm in
20. Output "B"
21. Wait for "three"
22. Arm out
23. Output "A"
24. Wait for "two"
25. Output "A" off
26. Slide left
27. Grip
28. Slide right
29. Arm in
30. Rotate left

Table 8.2. Continued

31. Ungrip
32. Output "C" on
33. Wait .2 seconds
34. Output "C" off
35. Wait for "reset"
36. Go to 1

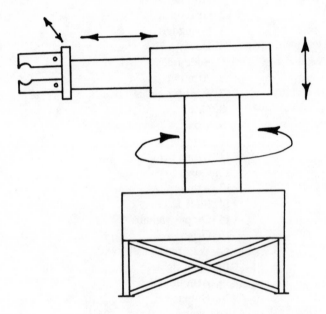

Figure 8.3.   The robot used to transfer parts between machines.

Table 8.3. Manual Vs. Robot

Manual

Parts Needed	100,000
Manual Time Average	37 + 63 = 100 sec.
Labor Costs	$8 hr. + 50% benefits = $12 hr. = .33¢/sec.
Manual Cost/Part	100 x .33 = 33¢/part
Seconds in Year at 8 Hrs./Day, 250 Days/Yr.	7,200,000
Manual Parts Produced	72,000

Robot

Robot Time Average	18 + 63 = 81 sec.
Parts Produced	7,200,000 ÷ 81 = 88,888
Robot Cost	$ 27,500
Robot Cost/Part	31¢
Robot Cost/Part, 2-Year Average	177,776 parts, 15 1/2¢

Table 8.4. Robot Interface

1. Open chuck	Command
2. Close chuck	Command
3. Part in gripper	Sensor
4. Part in rack	Sensor
5. Start machine	Command
6. Machine stopped	Sensor

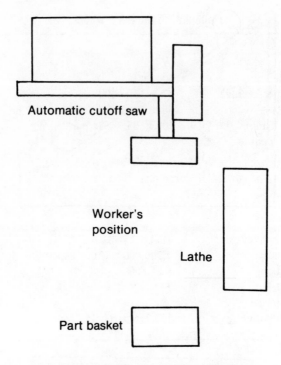

Automatic cutoff saw

Worker's
position

Lathe

Part basket

Figure 8.4.   The layout of manual machines.

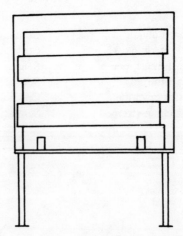

Figure 8.5.   A rack allowing for easy robot pick-up.

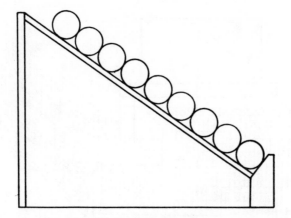

Figure 8.6. A self-feeding rack that holds bar stock before cutoff.

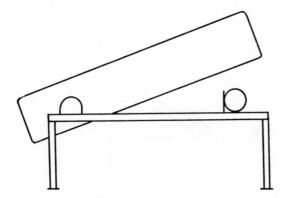

Figure 8.7. The cutoff saw showing end view of bar.

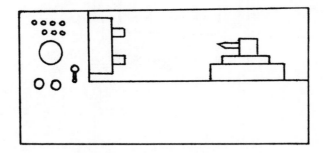

Figure 8.8. An automatic lathe.

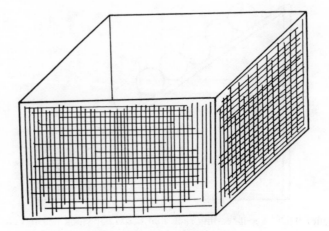

Figure 8.9.   The part basket.

Figure 8.10.   A cam-lever type limit switch.

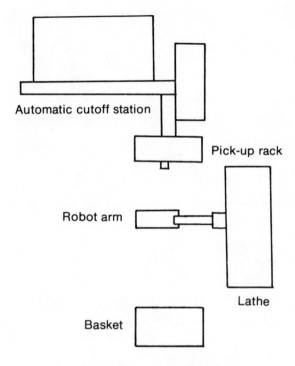

Figure 8.11.   Robot layout.

FOLLOW-UP

A follow-up study on this piece of equipment indicated that the time cycle was very close to engineers' estimates and that, as production went up, it was possible to run the machine an additional two hours per day at very little extra cost. The same foreman who oversaw the former operation and the setup man who previously installed new lathe tools and bandsaw blades now worked with the robot.

The cost savings to the customer calculated on a yearly basis were $25,533.33, assuming that 100,000 parts would be needed and produced on overtime, at a wage rate 1-1/2 times the normal rate. This further assumes that the robot would be used during the second year to produce the same number of parts at no additional cost.

CHAPTER 9

BROACHING OPERATION

The Acme Turnbuckle Company manufactures a bearing rod end. In their operation 7, two components are joined to make a turnbuckle end. Before the joining process, it is necessary to finish-machine one of the parts. The rough part is shown in Figure 9.1, the finished part in Figure 9.2.

To accomplish this finishing process, a broaching machine was used (Figure 9.3). A turntable on the broaching machine allowed one part to be loaded while the machine was working on another part. A finished part could be removed while the machining cycle continued. The joining process of the two components was done by a bearing press (Figure 9.4). The finished housing part was placed on the bearing press, and the bearing was placed in the correct position on top of it. The press was allowed to cycle and, by squeezing action, the bearing was forced into the correct position in the rod end housing.

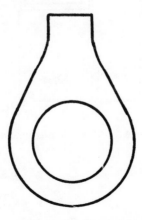

Figure 9.1. The rough form from which a turnbuckle is made.

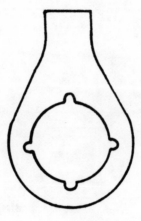

Figure 9.2.   Finished turnbuckle part.

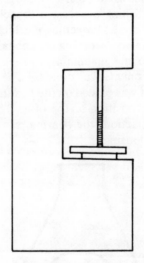

Figure 9.3.   A broaching machine.

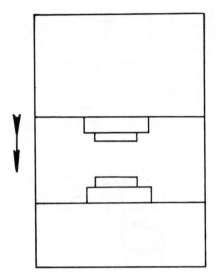

Figure 9.4. A bearing press.

Figure 9.5 shows a layout of the necessary equipment and supply baskets. A hook assembly on a traveling conveyer (Figure 9.6) transmitted the part from work station 7 to a dip process that protected the bearing component from corrosion during shipping and storage. The portion of the hook conveyer next to station 7 was power free—it did not have a drive mechanism. Rather, each time a hook contained a part, a small gate was released. By gravity feed, the hooked part proceeded through the rest of the plant. This arrangement allowed the machinery and the operator at station 7 to be unsynchronized with the rest of the plant.

INCREASED PRODUCTION NECESSARY

Sales for Acme turnbuckles had been very good. And, although the operator was working at very high efficiency and making 730 parts in an eight-hour shift, the plant was unable to meet the demand for the product. Production had to be increased, and the plant manager was in a quandry.

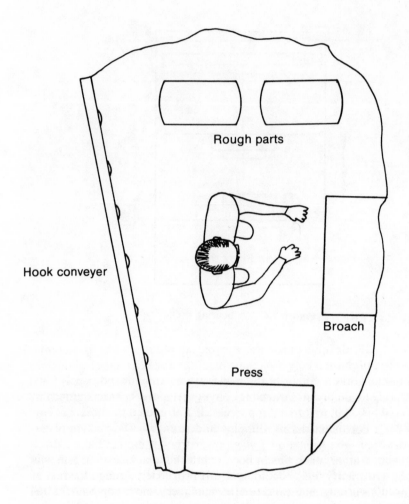

Figure 9.5.   The relative position of equipment and operator.

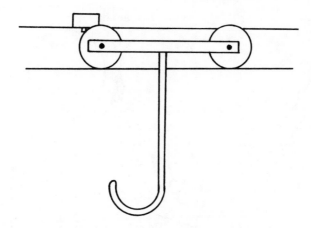

Figure 9.6.   A portion of the hook conveyer.

THREE ALTERNATIVES

He had to decide upon one of three viable alternatives. The first alternative was to purchase a new broaching machine and bearing press and hire an additional worker. Since the demand for parts was currently about 1, 300 per day, this might seem reasonable. However, a new broaching machine would cost $64,000, a new bearing press $31,000, and some additional baskets $2,800. An immediate capital expenditure of $97,800 would be necessary.

The second alternative was to use the existing machinery on a second shift. An additional worker could be hired and paid the same basic wage as the day-shift worker. At this particular plant, direct labor plus benefits and allowance for relief amounted to $12 per hour. However, since no other production was being done on a second shift, a foreman also would have to work the eight-hour night shift. State law and building security demanded it. The foreman's wages amounted to $14 per hour, which would add substantially to the cost of producing the part.

The third alternative for increasing productivity was to automate the process. Figure 9.7 shows a design layout using a robot with the existing bearing press and broaching machine. Additions with this automation package would be two spiral stacking conveyers (Figure 9.8) that would supply the parts to the robot.

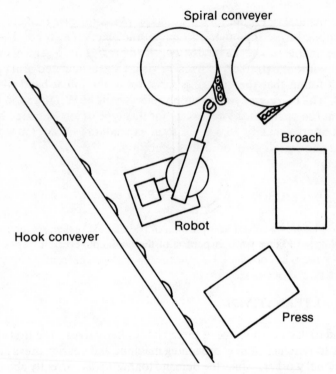

Figure 9.7.   The relative position of robot and equipment.

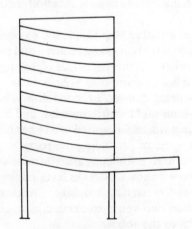

Figure 9.8.   A spiral conveyer.

In the manual operation, parts could be placed in wire baskets as they were supplied from the other portions of the plant. The robot, however, would be unable to select a randomly placed bearing or rod end blank from a basket without extremely expensive vision and articulated-arm options.

It was found that the cost of a robot to do the job at hand would be $31,400. The two conveyers to supply parts would be $7,150 and $6,800. A review of the connections necessary for this type of job indicated hookup costs of approximately $6,000. A total expenditure of $51,350 would be necessary to install a robot in this operation.

COST COMPARISON

Table 9.1. shows a cost comparison of the three alternatives open to the plant manager. The purchase of the new broach and new press would result, for the parts produced in addition to the ones currently being made, in a cost of 27¢ per part.

Table 9.1. Cost Comparison

Slowest portion of operation is the broach—39 seconds/part = 730 parts/day. Demand is approximately 1,300.

1. New Broach New Press		2. Second Shift		3. Robot	
Broach	$64,000	Wages & benefits		Robot	$31,400
Press	31,000	$12 hr.	$ 96 day	Conveyer	7,150
Baskets	2,800	Extra foreman		Conveyer	6,800
	$97,800	time	14 day[1]	Hookup	6,000
			$110		$51,350
54¢/part		15¢/part		2 shifts, total	
27¢/part if 2 years				yearly parts,	
				730,000 = 7¢/part[2]	

1. Part of foreman's time charged to other work
2. Does not include any maintenance

If a second shift were used, the foreman's wages would result in a cost per part during the second shift of 15¢.

The use of a robot, with a two-year buy-back, would result in a direct labor cost of 7¢ per part. This, compared with a cost of 13¢ per part for the human worker wages during the regular day shift, would result in a savings of 6¢ per part during the normal shift and 8¢ per part over an evening shift.

TECHNICAL FEASIBILITY

Previous experience by plant personnel indicated that a robot similar to the one shown in Figure 9.7 could perform all the functions of operation 7, providing the two conveyers were used. The conveyers could carry adequate supply to keep the robot running one entire shift without human intervention. It was planned, therefore, that the human operator would be transferred to other duties. As part of his duties, he would fill the conveyers once in the morning and once in the late afternoon just before the shift was over. Thus, the robot could run all evening until a malfunction occurred or until one of the conveyers ran out of parts. An automatic setup in its operating system would stop the robot when the parts were exhausted.

A request for quote was made to the robot manufacturer. A portion of the quote asked for details on the proposed conveyer hookup, allowing the robot to work without human intervention during the evening shift. The manufacturer proposed the following equipment (Figures 9.9–9.12): A gripper would sense the location of the rough part hole and would help avoid damage to the broach. An automatically resettable timer would indicate time lapse and receive a response from the robot signaling that all parts had been used or that some malfunction had occurred. Switching elements would sense the presence of parts on the conveyers, the broach, the bearing press and the conveyer hanger. Electrical connections already were available on the broach and on the part press to allow automatic cycling of the machines by the robot. A signal would have to be supplied to the conveyer to release a single hook loaded with its part, sending the part on to the next step.

The hookup connection chart shown in Table 9.2 indicates all of the connections necessary into or out of the robot. By using the switches at various places in the work stations as inputs, the robot would be able to duplicate the human decision-making functions as follows: Part present switches on the conveyers would indicate to the robot there were parts available and that the robot should continue. Part present switches on the press and broach would indicate to the robot that parts must be removed

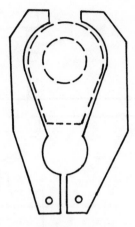

Figure 9.9. The gripping jaws of the robot.

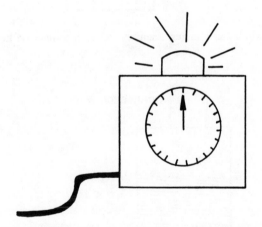

Figure 9.10. A timer to allow automatic signaling of a human operator.

before new parts could be placed in position. The part present switch indicating the presence of a hook would allow the machine to hang the rod end appropriately rather than merely releasing it in the air. The part present switch on the gripper could allow the robot to confirm, once its gripper was closed, that a part had been picked up.

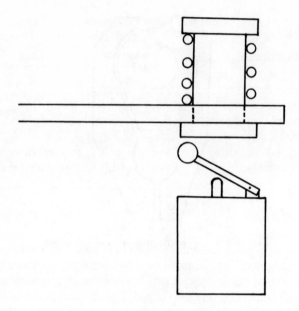

Figure 9.11.   Limit switch and plunger assembly to check turnbuckle holes.

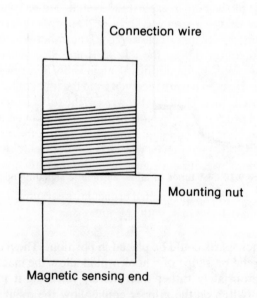

Figure 9.12.   A proximity switch.

Table 9.2. Robot Connections

Switch	Description	Normal State
1.	Part in conveyer	Open = no part
2.	Grip jaws closed beyond size of parts	Closed = no part
3.	Bearing in conveyer	Open = no part
4.	Hole OK in part	Closed, no hole in part makes switch open
5.	Hook ready	Open, closed if hook is ready for part

The probe and switch assembly shown in Figure 9.11 would be the most extensive "judgment" mechanism used by the robot. Because of the way the broach carried out its cutting process, the rough turnbuckle end must have a hole of approximately the right size, shape and position. To ensure that the hole was correct—since no human operator would be there to examine it—the robot would place the rough turnbuckle over the probe and push down. If the hole were too small or in the wrong position, the probe would be pushed down against the switch. An *on* switch would indicate to the robot that this part was to be scrapped and placed in a separate bin.

The auto reset timer connected to the robot would be used to signal a human operator during the day shift and stop the robot when its parts supply was exhausted during the night shift. The timer would be set for 45 seconds. If a signal were received from the timer indicating that 45 seconds had elapsed since it had last been reset, the robot would cause a jump inside the robot program to a specific routine. The robot controller would turn on a flashing light indicating that there had been some malfunction or that the robot had stopped doing its work. The robot would then turn off the rest of its functions, awaiting a signal from the operator that work was ready to begin.

Table 9.3 shows the manual broaching operation; Table 9.4, the program to accomplish the task with robotics.

SELECTION PROCESS

The robot selection process was a relatively simple one. Since the engineers at the company were already familiar with the robot under consideration, they felt more comfortable with it and preferred it to untried equipment. A review of the marketplace indicated that there probably was no robot with the same features that was substantially less expensive.

The robot selection process included two important criteria. First, because the factory was already behind schedule, any retrofitting must be accomplished very quickly to lose as little productivity as possible. Second, the factory's parts must be produced at the same or lower cost. Although the parts were selling very well, it was believed any price increase might adversely affect sales. The robot under consideration had a reputation, based on plant experience, of being very fast to install. What's more, no parts price increase would be necessary. The robot choice was easy.

Table 9.3. Manual Operation

1. Pick up rough part
2. Check hole diameter
3. Place in broach table
4. Pick up finished part
5. Close cover on broach
6. Place part in press
7. Pick up bearing
8. Place bearing in press
9. Remove hands
10. Cycle press
11. Remove parts
12. Place finished unit on hook
13. Release hook

Table 9.4. Robot Program

1. Arm in
2. Raise to level 2
3. Wait for conveyer switch 1 closed
4. Output 1 reset timer
5. Rotate to rough part rotation
6. Extend to rough part extension
7. Down to level 1
8. Grip
9. If switch 2 closed, wait
10. Up to level 3
11. In full
12. Rotate to broach drop-off
13. Extend to broach drop-off
14. Down to level 2
15. If switch 4 open, wait
16. Release
17. Up to level 3
18. Grip
19. If switch 2 open, wait
20. In full
21. Release
22. Rotate to broach pick-up
23. Extend to broach pick-up position
24. Down to level 2
25. Grip
26. If switch 2 is closed, wait
27. Up to level 3
28. In full
29. Rotate to press
30. Extend to press position
31. Down to level 2
32. Release
33. Up to level 3
34. Grip
35. If switch 2 is open, wait
36. In full
37. Rotate to bearing conveyer
38. Release
39. Extend to bearing conveyer
40. Down to level 1

Table 9.4. Continued

41. Grip
42. If switch 2 is closed, wait
43. Up to level 3
44. In full
45. Rotate to press
46. Extend to press
47. Down to level 2
48. Release
49. Up to level 3
50. Grip
51. If switch 2 is open, wait
52. In full
53. Output to press on
54. If switch 6 is open, wait
55. Extend to press
56. Release
57. Down to level 2
58. Grip
59. If switch 5 is open, wait
60. Up to level 4
61. If switch 2 is open, wait
62. Rotate to hook
63. Down to level 3
64. Release
65. Down to level 2
66. Grip if switch 2 is open, wait
67. Output to release gate
68. Wait 5 seconds
69. If switch 5 is closed, wait
70. Go to 1

INSTALLATION

The robot installation was planned to begin at the end of the shift on Friday afternoon. It was hoped that three days would be sufficient. The estimate was that by the end of Tuesday evening's run, production would be restored to its previous level.

When the actual installation took place, it was found that careful planning of the input and output signals and adequate preparation of the conveyers and switching mechanisms allowed for faster hookup than expected.

A follow-up study of the system indicated that its reliability was very close to that predicted. An average of only one malfunction in two months occurred, such as a jammed part during the evening shift. Overall operating efficiency for the system, based on actual versus projected production, was approximately 98 percent during the first year.

CHAPTER 10

PALLETIZING ENGINE BLOCKS

High Performance Motors produces components for the auto racing and recreational off-the-road vehicle market. At one plant, the final operation was to palletize the finished machined engine block so that it could be shipped to another plant for assembly. This manual operation was a bottleneck to high productivity.

Figure 10.1 shows an operator next to the power-free conveyer at the end of the line. Two such people worked unloading engine blocks from the conveyer and placing them on pallets. Pallets were placed in two different positions so that while blocks were being loaded on one, a lift truck could remove a full pallet. To assist in the unloading process, there was a trolley-type overhead crane with an electric lift motor and an air-operated gripping device that could be positioned by the operator.

Each of the operators took an engine block, lifted it with the crane mechanism and placed it on a pallet. After six engines were on a pallet, a piece of cardboard was placed on top of the engines. Then, another layer of six engines was added to the pallet. Factory machinery produced one engine block every 22 seconds. Due to the time constraints of the lift-and-carry mechanism, two operators were needed, each handling every second engine.

THE PROBLEMS

It was determined that over a year, an average of eight engines per day were damaged, some by an operator dropping the finished component, others by a worker knocking an engine against others on the pallet. Still others were damaged in shipping because of improper placement of the blocks on the pallet.

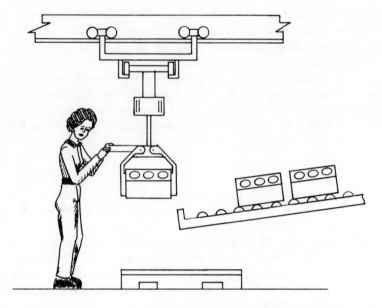

Figure 10.1.   The manual unloading of engine blocks.

Studies showed that six minutes per day were lost as a direct result of this palletizing operation. As pallets became full or engine blocks were in the wrong position and had to be straightened, blocks accumulated in the power-free conveyer. If more than 20 blocks stacked up in the conveyer, a blockage to the last machine process developed, and all of the preceding machining operations also stopped. Efforts to design the machinery in a way that would allow for greater stackup of unpalletized parts had been unsuccessful.

COSTS

An analysis of costs provided the following information: The direct labor rate at the plant was $14 per hour. However, with the addition of fringe benefits, relief time and overhead for supervision, the labor rate became $26 per person-hour. Production was run at only 7-1/4 hours per day, so that the total number of parts produced in an 8-hour shift was 1,227. The value of the part used for accounting purposes at the factory was $74. The raw material price was $27.50 per part, so the value of lost production might have been considered to be $46.50 per part. However, because the damaged parts had no substantial salvage value, their value was computed at $74 per part.

Since an average of eight parts per day were scrapped, the economic incentive to improve the loading process was $592 per day. Six minutes per day of lost production time meant that 16.36 engines were not produced, for a lost value of $760 per day. The total indirect cost for the palletizing process was $1,768 per day (Table 10.1).

Table 10.1. Three Factors of Loss

1. Workers' time spent loading
   2 workers @ $26 each x 8 = $416/day

2. Engines never made due to stopping of line

   $$\frac{(6 \times 60)}{22 \text{ sec.}} \times (\$74.00 - \$27.50) = \$760/day$$

3. Engines scrapped due to damage
   8 engines x $74 = $592/day, finished product including raw material

   Total $1,768/day

ROBOT CONSIDERED

In determining how to improve the loading operation, a robot was considered. Figure 10.2 shows a potential robotic conversion. Based on phone conversations with several robot vendors, it was estimated that a robot to do this job would be quite large and of the hydraulic type. Prices for the base equipment were all about $92,000, so this number was used to calculate the potential for savings.

If we assume that the robot would not drop any engines, once it was programmed properly, and that it would be able to palletize the engines continuously at a rate of 22 seconds each, the payback on the robot for a two-year period would be $184 per day. By subtracting this number from the $1,768 per-day loss with the manual process, there was a potential per-day savings of $1,584.

Each minute of lost production time had a cost of $126.82. If we divide the $1,584 per day potential savings by this amount, we get 12.49 minutes as the time we could tolerate the robot not running each day and still have it equal the cost effectiveness of the manual operation. This would mean that if the robot operated only at 97.4 percent uptime, it could still be considered an economic alternative (Table 10.2).

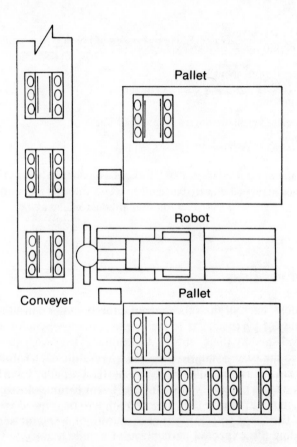

Figure 10.2.   Layout of robot unloading station.

Table 10.2. Robot Vs. Manual

1. Payback
$$\frac{\$92,000 \text{ cost of robot}}{500 \text{ days' work in 2 years}} = \$184/\text{day}$$

2. Lost production value
$$\frac{60 \text{ sec./min.}}{22 \text{ sec./part}} = 2.7272 \text{ engines/min. x } \$46.50 = \$126.82/\text{min.}$$

3. $$\frac{\$1,584 \text{ /day savings}}{\$126.82/\text{min. cost of lost production}} = 12.49 \text{ min./day equivalent cost of production}$$

4. $$\frac{12.49 \text{ acceptable time loss/day}}{480 \text{ min./day}^1} = 0.026$$

100% - 2.6% = 97.4%

1. 8 hours (480 min.) is used instead of 7 1/4 actual production hours (435 min.) to allow a more realistic comparison to other tables showing robot uptime.

SELECTION PROCESS

Upon request, each of the robot manufacturers under consideration offered examples of tasks similar to the engine-loading operation that were easily accomplished by their robots. Additionally, they each quoted items to be added to the basic system. One was a hydraulic obstruction (Figure 10.3) to prevent the robot from entering a particular pallet's working area while a lift truck was present. Another was a simple beam detector to signal the robot when a part was in the proper position to be gripped and loaded.

As part of the robot selection process, questions were put to manufacturers regarding the expected percentage of uptime attainable with their robots. The choice was narrowed to two potential vendors, each of whom indicated that its robot was rated at 98 percent uptime. One of the vendors had the advantage of a local service organization with resident staff and automatic diagnostic equipment. It also offered an employee training program free with the purchase of a robot. This vendor was selected, even though the initial robot cost was some $2,000 more than that of other manufacturers' models.

INSTALLATION

Installation of the robot required little factory downtime. While the plant was running, the robot was positioned on its base in front of the conveyer. During the nonproductive shifts, the robot connections were made and the robot program established.

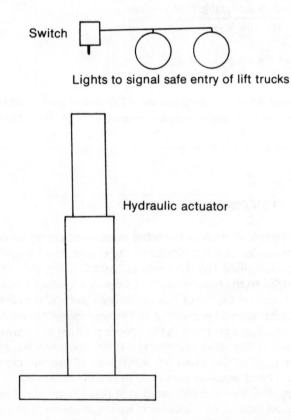

Figure 10.3.   A hydraulic safety ram to restrict motion of the robot.

FOLLOW-UP

A follow-up study of the robot after six months of operation indicated that, indeed, a high percentage of uptime had been achieved. During the six months, the robot had caused a loss of productive time 17 times, but with an average downtime of only 2.5 minutes. Once, the robot had malfunctioned and required the replacement of a component—downtime was 31 minutes. This amounted to 73.5 minutes of lost production time.

In comparison to an expected downtime with the manual operation of 750 minutes (6 min./day x 125 working days), a saving of 676.5 minutes of production time was achieved—a value to the company of $85,793.73 (676.5 x $126.82) or very nearly the price of the robot itself. During the study period, no engine block had been damaged due to the loading process.

CHAPTER 11

WELDING

Iron Pumpers of America is a manufacturer of heavy-duty weight benches for the professional spa and gym market. A portion of the weight bench is a metal pole with a yoke welded on one end that will support half of a barbell (Figure 11.1). Several holes are placed along the metal shaft to allow for proper positioning.

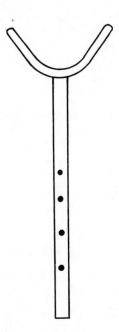

Figure 11.1. A welded metal yoke.

In making yokes, parts had to be positioned manually, clamped in place and welded by hand. This operation required 30 seconds for each yoke produced.

ROBOT FEASIBILITY

Because of increased business and a desire for increased production, an engineering study was initiated to test the feasibility of using a robot. A review of the available technology indicated that the parts probably still would have to be manually loaded but could be welded satisfactorily by a robot holding a welding gun (Figure 11.2).

After checking with potential vendors, engineers determined that automatic clamping equipment, to be loaded by an operator, and an industrial robot would cost approximately $90,000 (Figure 11.3). It was es-

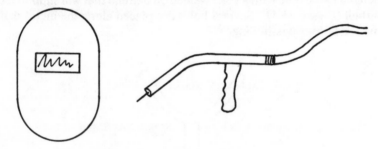

Figure 11.2.   Manual welding equipment.

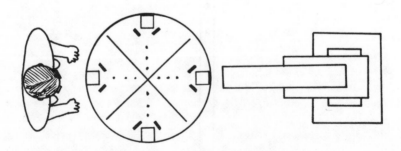

Figure 11.3.   The robot welder/manual loading station.

timated, based on the inches of weld and motions involved, that a part could be produced in 12 seconds.

COST COMPARISON

An analysis of the cost to produce each part manually versus by robot is shown in Table 11.1. Total wages and benefits for this particular plant amounted to $20 per human worker hour. At a production rate of 30 seconds for each part, a cost for the value added was 16.6¢ (20 ÷ 120). The cost of a combination of a robot welder and a human loader could be calculated based on a two-year payback for the robot. At a price of $90,000, the robot would cost $22.50 per hour (90,000/4,000 hrs.). The cost of the human worker still would be $20 per hour, for a total per-hour cost of $42.50. As one part would be made in 12 seconds, the cost per part for the welding process would be 14.1¢ (42.50 ÷ [60 x 5]). Clearly, there was little economic incentive for the use of a robot. The engineering study was discontinued, and plans were made to install another manual operation in the factory for the welding of yokes.

Table 11.1. Manual Vs. Robot

Manual Weld/Manual Load	Robot Weld/Manual Load
30 sec./part	12 sec./part
$20/hr.	$22.50/hr. (2 years)
16.6 ¢ /part	+20.00/hr.
	$ 42.50 /hr.
	14.1¢/part

The company considered what would have happened had the robot been installed and had demand fallen so that one person could have done the welding. A loss would have been sustained by the company, unless the robot could have been resold at a major portion of the initial price. On the other hand, by installing a second manual installation, the new employee could be released should product demand fall. And, no additional substantial labor cost would be incurred.

A NEW ELEMENT

Before a second manual operation was installed, however, a new element entered into the decision-making process: product liability. The company's yokes had, in several instances, broken apart while in use. While no injury had been suffered, the insurance carrier was concerned that one major accident would initiate a substantial lawsuit against Iron Pumpers of America. At the suggestion of the plant foreman, a study was initiated by the insurance company as to potentially lower rates should the yoke be welded by robot and therefore be of more consistent and higher quality. The insurance company found that with robot welding, they could lower Iron Pumper's premium by $31,000. Robotic feasibility was again studied.

VENDORS CONTACTED

Upon request, several robot vendors offered to demonstrate the high-quality welding possible with an industrial robot. Of the 20 robot vendors considered for the robot conversion, four responded that they would be willing to take total system responsibility for the automation process. Management requested that these four vendors bid on automating the yoke-weld process. Table 11.2 gives an analysis of their quotes. The prices shown include the robot itself, the turntable equipment used to clamp and transport the yokes, the first robot program, hookup of all the equipment and trouble-shooting at the time of initial use.

All four vendors offered to train plant personnel in the operation and maintenance of the equipment, but only two had local trainers who would

Table 11.2. Robot Conversion Analysis

Total package bid: robot, turntable, first program, hookup, trouble-shooting, training

	Vendor 1	Vendor 2	Vendor 3	Vendor 4
Price	$87,400	$91,260	$90,000	$104,200
Training	+	+	–	–
Local Service	+	+	–	–

be available after the system was initiated. Based on a quote of $87,400, a robot was purchased.

TRAINING

Once the robot was installed, a training program was initiated. Several operators were trained to load parts and work with the robot. They learned how to adjust welding nozzles, how to identify faulty welds due to mispositioning of the robot and how to install new rolls of wire. Four hours of training was provided each of the operators. Training was also provided in maintenance. Three electricians were each given a three-day course in how to make simple repairs and how to routinely maintain the robot for long-life operation.

FOLLOW-UP

A study done after seven months of operation found there had been no downtime attributable to robot malfunction. There had been one problem with the type of wire used in the welding gun, and 79 parts had had to be scrapped before the problem was discovered. However, at the time of the study, the robot was functioning adequately and making parts above the quality standard established by the insurance company and Iron Pumpers of America.

CHAPTER 12

CHAIR GLUING

Transworld Plastics manufactures a line of office lobby seating. The seat construction involves the gluing of two plastic halves. Figure 12.1 shows where the glue must be applied. The company wanted to change its gluing process to use a hot-melt-type glue. The objectives were to get better bonding of its chairs and, because less glue would be necessary, to save money.

THE PROBLEMS

There were two technical problems. One was that the glue had to be dispensed in exactly the right position to minimize use of the glue. The other was that the cure time for hot-melt adhesive is very short. The adhesive manufacturer indicated that from the time the first bit of glue was applied to the chair, no more than 24 seconds could elapse before the joining part had to be in place.

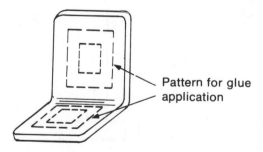

Pattern for glue application

Figure 12.1. A seat section showing the position of glue.

An analysis of the length of the glue lines and the needed accuracy indicated to Transworld Plastics that this operation was not feasible if done by human hands. The company also found that a machine dedicated solely to the process of dispensing glue along the required path would be extremely expensive and would be difficult to change when Transworld began to manufacture other styles of seats.

ROBOT FEASIBILITY

Robots were considered as an alternative. Figure 12.2 shows the basic type Transworld's engineers proposed using—a small, fast, and accurate electric robot.

Transworld's engineers determined that 3 seconds would be necessary for a conveyer to move the seat sections into position. Then, the robot, to do its gluing job, would need a maximum of 20.5 seconds, as follows: One-half second would be allowed for the robot to enter the working range of the seat. Moving at a constant velocity of eight inches per second, the glue gun would glue the required 152 inches in 19 seconds. One-half second would be allowed for switching between individual paths and one-half second for exiting the seat area.

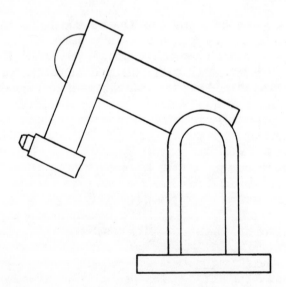

Figure 12.2. The robot to be used for glue application.

Four seconds would be required for another device to press the chair seat to the chair back. Next would come 15 seconds of cure time, during which the gluing robot would remain motionless outside the working area of the chair. After this sequence, the glued chair would be removed and new seat sections would be moved into position.

The same automated device used to join the seat sections would be used before the gluing operation to unload the conveyer. It would also be used to stack finished chairs in shipping containers.

VENDOR QUOTES

Transworld engineers asked five vendors to quote a robot suitable for this operation and determine the robot's ability to meet the time constraints. These five vendors were selected based on the following considerations:

1. Each was on an approved vendor list supplied by Transworld Plastics' parent corporation.
2. Each had a local sales and application office.
3. Each had a robot with the necessary accuracy.
4. Each had a robot that could be located safely out of the way of the conveyer and still could reach easily all of the required points.
5. Each had a robot with sufficient weight-handling capabilities to carry the loaded glue gun.
6. Each responded to a request for information by Transworld.

Table 12.1 shows a comparison of data submitted. A plus in a vendor column indicates a positive factor; a minus, a negative factor; a zero, not a factor. We can see that robot prices were somewhat similar, as were robot accuracies and speeds in the standard test. Although the selected robot would be allowed a full 20.5 seconds during the dispensing cycle, each of the proposed robots was capable of traveling faster than that if necessary. Each was also capable of receiving the input signal from the conveyer indicating a seat section was in place and ready for gluing.

Since Transworld Plastics would be providing the glue gun, that item would not enter into the selection process. Figure 12.3 shows the glue nozzle that would be used with the robot. It would contain a supply tube from which a flexible yet solid rod of hot-melt glue would be forced into the nozzle. A power connection would supply electricity to the heating chamber. Figure 12.4 shows how the gun would be mounted to the robot wrist rotation disk at a 90° angle, thus allowing the nozzle to dispense adhesive horizontally or vertically.

Table 12.1. Five-Vendor Comparison

	Vendor 1	Vendor 2	Vendor 3	Vendor 4	Vendor 5
Base Price of Robot	$58,000	$61,000	$54,000	$52,000	$57,500
History in Glue	+	+	-	o	o
First Program Free	+	+	+	+	+
Robot Time Trial (in min.)	13	15	14	15	13
Robot Accuracy	±.008	±.006	±.010	±.010	±.008
Local Training	+	-	+	-	-
Ease in Programming	+	-	+	+	o
Diagnostics	+	+	+	+	o

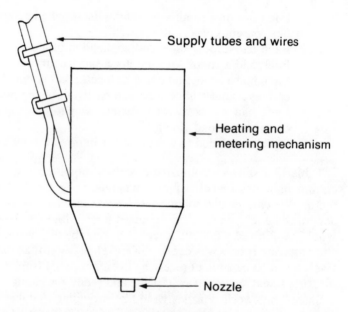

Supply tubes and wires

Heating and metering mechanism

Nozzle

Figure 12.3.   A glue-delivery system.

SELECTION

The vendor selected by Transworld was Vendor 3. Although its robot was not the lowest priced, it had a feature important to Transworld that overshadowed the additional cost of $2,000 over the lowest bidder; specifically, the ability to do local training for service people.

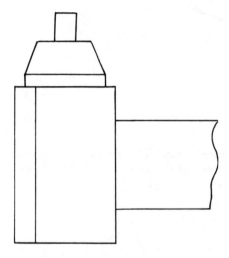

Figure 12.4.   The glue gun mounted on the end of the robot.

When the robot was installed, only two days were necessary to connect the robot, put it in position and write the initial program for the gluing process. The robot was readily adaptable to the glue nozzle equipment, and little effort was necessary to make the entire system a functional package.

When the robot was first run, however, there was a problem with glue *wipeoff*. Figure 12.5 shows an example of glue wipeoff, a phenomenon characteristic of any viscous material forced through an orifice and into a bead shape. Indeed, soft ice cream and taffy exhibit this same phenomenon. A minor change in the program was sufficient to prevent the misplacement of glue and involved a quick oscillation to part the adhesive substance (Figure 12.6).

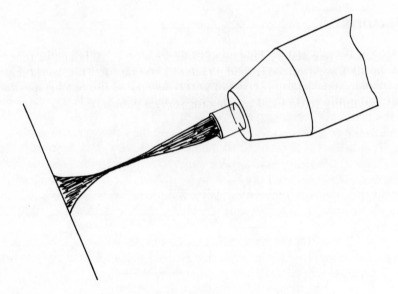

Figure 12.5. The glue has a tendency to form long strings at the end of the process.

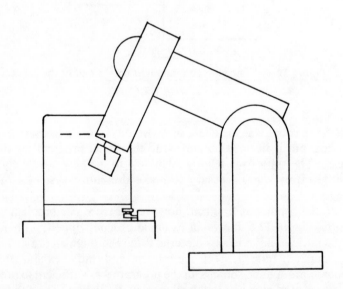

Figure 12.6. Dispensing of glue with a robot.

FOLLOW-UP

After sufficient production had been made to do a worthwhile study, it was discovered that a significant number of seats were being misglued by the robot. An analysis of this problem indicated that the robot was not at fault; rather the problem was caused by the mispositioning of the seat itself on the conveyer locating pins.

When the seat bottom was first installed on the conveyer, it was placed over four pins that fit into the support brackets later used to support the chair legs. These pins were small enough to fit easily within the chair leg openings. However, at a previous work station, the trim operation often moved the chair out of position. Since the locating pins were not large enough to keep the chair bottom in place, a misapplication of the glue resulted.

Figure 12.7 shows the correction that was made— tapered locating pins were used to properly position the seat prior to gluing and trimming. After this correction, there were few malfunctions.

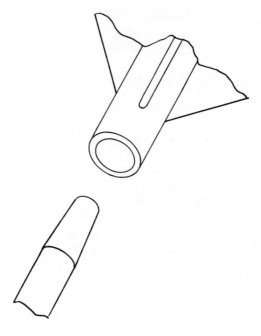

Figure 12.7.   A tapered pin locates plastic leg holder exactly.

An analysis of the robot uptime indicated that after three months, the robot had been operating 94 percent of the time. Three percent of the downtime was attributed to malfunctions of the robot or its control cycle. Two percent was caused by nozzle malfunctions such as clogs or burned out heating elements. One percent was caused by the location switch on the conveyer. Table 12.2 shows a sample program created to run the gluing operation.

Table 12.2. Program for Gluing

1. Wait for signal from switch
2. Move to part
3. Move to exact location, position 1
4. Energize heater
5. Wait 0.2 seconds
6. Start dispense cycle
7. Position 2
8. Position 3
9. Position 4

Many more positions

68. Stop dispense
69. Go to inner box
70. Start dispense

Many more positions

274. Stop dispense
275. Stop heater
276. Away from part
277. Back to start position
280. Signal ready for joining
290. Wait for switch equal to off
300. End of program

CHAPTER 13

CHAIR POSITIONING AND STACKING

Transworld Plastics also determined that the seat-making operation had to be automated both before and after the gluing process. The additional automation would have to be carried out within the working range of the gluing robot, but at no risk to human safety. To accomplish this, company engineers anticipated the use of an additional robot for positioning and stacking.

SECOND ROBOT

The second robot would lift a seat back from a supply container and carry it to the glue station. At the appropriate time, the robot would place the seat back onto the glued lower section. It would wait, then pick up the finished seat, place it in a shipping container, get a piece of cardboard from a supply container and place it on top of the seat (Figures 13.1–13.3). Then it would begin the cycle again.

Other equipment that would be used with the robot is shown in Figures 13.4 – 13.6. A specially designed gripper that would allow a seat to be picked up by two vacuum cups was designed by Transworld engineers and was adaptable to virtually any robot. The gripper, when joined to the chair, had a weight of 29 pounds, so the proposed robot had to be able to handle this weight. The two vacuum cups produced vacuum exactly where it was needed. Air lines and valving could be mounted on the robot body and the vacuum controlled at will.

Because of the operation of the vacuum venturi, a certain amount of air pressure would produce a vacuum at the cup. But, a greater air pressure would overpower the venturi and force air out through the cup. This would allow the gripping unit alternately to grasp and to release a chair section. A

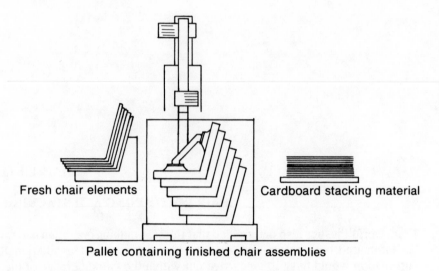

Fresh chair elements

Cardboard stacking material

Pallet containing finished chair assemblies

Figure 13.1.   Robot loading of finished chair assemblies.

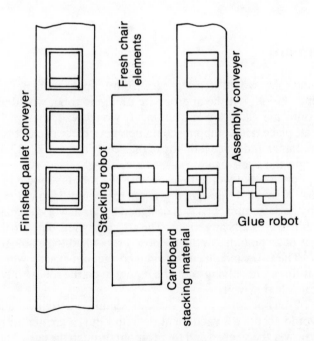

Figure 13.2.   The two-robot chain-assembly station.

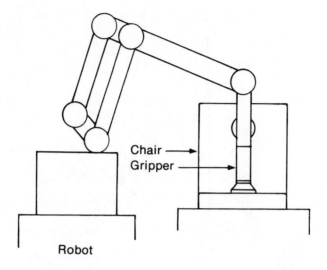

Figure 13.3.   A side view of the stacking robot.

vacuum-operated switch would be attached to each of the venturi mechanisms where they were joined to the vacuum cup. Each switch could measure the level of vacuum its cup was using and signal the robot whether a part was in the gripper properly or whether a part was absent.

Electrical limit switches, activated mechanically, would be used on conveyers to indicate that a seat bottom was in the right position for placement of the upper seat and that a shipping container was in the proper position for filling.

Table 13.1 gives an example of a program planned to implement this robot task. Written in general terms, the program would be adaptable to any of the proposed robots.

VENDOR QUOTES

A sample of this program was sent with the request for quote to each proposed robot vendor to obtain its assurance that its robot could render all of the functions necessary to program. Table 13.2 indicates the output and input signals necessary for the operation of the robot.

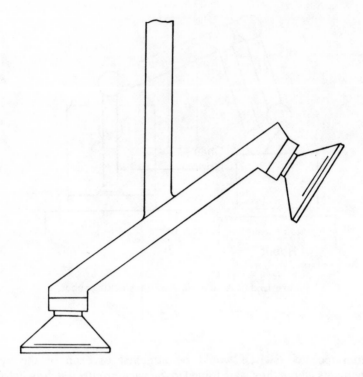

Figure 13.4.   A two-cup vacuum gripper for lifting chairs and components.

Figure 13.5.   A vacuum cup.

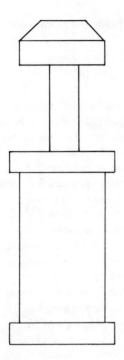

Figure 13.6 An air cylinder fitted with a rubber "snubber" end-piece.

Table 13.1. Program For Seat Joining And Stacking

Main Program

   1. Swing to seat stack
   2. Vacuum on
   3. Down until interrupt
   4. Reset counter A
   5. If switch 1 is on, jump to 10
   6. If counter A equals 3, jump to 50
   7. Wait 0.4 seconds
   8. Add 1 to counter A
   9. Jump to 5
  10. If switch 2 is on, jump to 20
  11. If counter A equals 3, jump to 50

Table 13.1. Continued

12. Wait 0.4 seconds
13. Jump to 10
20. Up
21. Swing to ready position
22. Signal glue robot "OK to start"
23. Wait for 'glue finished'signal
24. Move to position 1
25. Move to position 2
26. Move to position 3   } Path to join position
27. Move to position 4
28. Move to position 5
29. Wait 15 seconds
31. If switch 1 is off, jump to 50
32. If switch 2 is off, jump to 50
33. Move up
34. Signal "OK for new seat"
35. Swing to shipping box
36. Go to subprogram No. 1
37. Go to subprogram No. 2
38. If counter B equals 5, jump to 40
39. Jump to 1
40. Signal new shipping container
41. Wait until switch 3 equals 0
42. Wait until switch 3 equals 1
43. Counter B equals 0
44. Jump to 1
50. Signal operator
51. Wait for reset from operator
52. Jump to 1

Subprogram No.1

1. Jump to 10 if counter B equals 0
2. Jump to 20 if counter B equals 1
3. Jump to 30 if counter B equals 2
4. Jump to 40 if counter B equals 3
5. Jump to 50 if counter B equals 4
10. Down to level 1
11. Jump to 111
20. Down to level 2

Table 13.1. Continued

21. Jump to 111
30. Down to level 3
31. Jump to 111
40. Down to level 4
41. Jump to 111
50. Down to level 5
51. Jump to 111
111. Blow off vacuum
112. Up 6 inches
113. Blow off stop
114. Vacuum on
115. If switch 1 is on, go to 50
116. If switch 2 is on, go to 50
117. Up above box
118. Return to calling program

Subprogram No. 2

1. Swing over paper
2. Down until interupt
3. If switch 1 equals 0, go to 50
4. Up
5. Swing over box
6. Call subprogram No. 1
7. Add 1 to counter B
8. Return to calling program

All of the proposed vendors were given the following robot requirements: The robot must have the ability to operate a subprogram from the
main program. It must be able to provide an interrupt signal to stop robot
motions in progress. It must be able to jump to a previously assigned position and go through assigned motions as a part of another program. It must
have the ability to swing on its base 300° of arc. It must have the ability to
travel 49 inches in a vertical direction while at the same time travelling 61
inches in a horizontal direction, with an accuracy of .010 inch or better. It
must be able to lift 29 pounds throughout the working robot range.

Table 13.2. Output and Input Signals

Outputs	To	Message
1	Glue robot	OK to glue
2	Shipping conveyer	New box needed
3	Glue conveyer	OK for new seat
4	Upper chair conveyer	Need more parts

Inputs	Message
Switch 1 = 0	Cup 1 has less than 29 in. of vacuum
Switch 1 = 1	Cup 1 has more than 29 in. of vacuum
Switch 2 = 1	Cup 2 has more than 29 in. of vacuum
Switch 2 = 0	Cup 2 has less than 29 in. of vacuum
Switch 3 = 1	Shipping box in place

In all, eight vendors were able to satisfy the above requirements. Of these, two vendors had hydraulic-type robots, and the other six had all-electric robots. Because of previous experience with hydraulic units in another branch of the company, it was decided that a hydraulic robot was inappropriate. The great strength of a hydraulic robot was not needed, and an additional cost of several thousand dollars per year to produce the hydraulic pressure would be necessary.

Table 13.3 shows a breakdown of information provided by the six electric robot quotes. We see that Vendor 6, the lowest bidder, did not have either local service or training, or the ability to interact directly with the computer. (It was determined that this ability would be a plus in the future, even though it was not necessary for the task at hand.) What's more, Vendor 6 was not approved by the parent organization of Transworld Plastics. The next lowest priced vendors, Vendor 2 and Vendor 5, had local training and service and computer interaction ability but were not approved by the larger organization.

Table 13.3. Robot Vendors

	Vendor 1	Vendor 2	Vendor 3	Vendor 4	Vendor 5	Vendor 6
Price	$91,600	$84,700	$88,020	$90,107	$87,165	$82,000
Local Service	+	+	+	+	+	-
Local Training	+	+	+	-	+	-
Computer Interaction Ability	+	+	+	-	+	-
Approved	+	-	+	+	-	-

All

Subprogram Ability
Signal Interrupt
Conditional Branch
300° Swing Base
49 in. Vertical Travel, 61 in. Horizontal
Accuracy of 0.010 In. or Better
29 Lb. Lift

SELECTION

A preliminary decision was made to select Vendor 3, based on a combination of its relatively low price and its strength in the other important areas.

Before the final decision was made, however, a check was made with the parent company research and development department to see why Vendor 2 had never been approved for use. It was discovered Vendor 2's robot was under test and had been performing admirably. In fact, a report was being drafted, indicating that this robot should be approved for use. The decision was made that Vendor 2's robot would be used.

Once the robot was installed, a safety shield was added to the plant area. Figure 13.7 shows how this shield was placed around the working area of the two robots to prevent incidental entering and potential danger to human workers.

FOLLOW-UP

An analysis of the robot after four months of operation showed that 98 percent of the time it was working properly. The major reason for downtime was a failure in the vacuum-sensing mechanisms attached to the gripper. At times, when the gripper had attached itself to a cardboard insert, enough air had leaked past the cardboard to falsely trigger the vacuum-sensing device. Since the vacuum sensor would indicate less than 29 inches of vacuum, the device would signal in step 50 of its program that an operator should come to attend.

To correct the problems, a change was made whereby the switch would activate at 27 1/2 inches of vacuum. After that, there were no malfunctions attributable to mistriggering of the switches. Even at 27 1/2 inches of vacuum, no instances were discovered where a seat fell out of the vacuum cups or was mispositioned due to a low level of vacuum.

Another reason for downtime was burnout of one axis driving unit in the robot. Because of robot modular construction and because plant personnel had been trained adequately, only 13 minutes of production time were lost in the diagnosis and replacement of the unit.

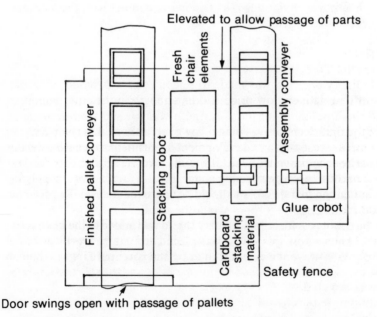

Figure 13.7.   A safety barrier keeps unsuspecting onlookers away from the robot.

CHAPTER 14

SPRAY PAINTING

The Wondercut Lawn Equipment Company manufactures a variety of home and commercial lawn mowers. The need for Wondercut to use robotic automation became evident after a study was conducted involving the safety of spray painters in the factory. A letter to management from the union representing Wondercut workers detailed the study results. The study indicated the repeated inhaling of paint vapors had been harmful to workers. Moreover, paint deposits coming in contact with the workers' skin also presented a health risk.

ROBOTIZING THE JOB

In a spirit of cooperation, the workers, the union and the company agreed to the automation of the spraying operation and the retraining of painters to program robots and carry out other jobs within the plant.

The process of robotizing a spray-painting operation can be one of the simplest in the robot industry. And, although Wondercut engineers had no previous robotic experience, contacts with companies supplying spray-painting equipment demonstrated to them the task easily could be accomplished.

There were three companies nearby that could provide the necessary equipment (Table 14.1). Vendor 3 was eliminated at the outset for being overpriced. Vendors 1 and 2 appeared to be almost identical in equipment and features. Since Vendor 1 had the lowest price and the shortest delivery time, it was selected.

The equipment purchased for the spray-painting job was a hydraulic robot of the teach type. It is programmed by manually guiding the tip of the robot through a series of motions. Then, those motions can be duplicated by the robot whenever desired.

Table 14.1. Vendor Comparison

	Vendor 1	Vendor 2	Vendor 3
Price	$28,600	$31,250	$79,470
Teach Type Programming	+	+	+
Hydraulic	+	+	+
Acceptable Accuracy	+	+	-
Local Service	+	+	-
Local Training	+	+	-
Delivery	2 days	2 weeks	

Because of the ease of installation and programming, the robot was installed in one evening when production was not running, and a test program was available for use with the following day's production. Figure 14.1 shows the part —the body of a small tractor—to be spray painted. Figure 14.2 shows the robot's path to carry out the spray-painting process.

When the robot was first installed, a curtain available at the plant was placed around the spray-paint operation. It was discovered, however, that the curtain was insufficient to stop paint vapors from entering the other areas of the plant. So, a booth with solid walls and ceiling was installed around the robot equipment (Figure 14.3). Inside the booth was placed a large fan that continuously exhausted air, passing it through a filter. The filter stopped the escape of paint particles. A constant influx of air through the booth's doorway and the openings through which parts passed prevented spray paint from exiting the booth at these locations.

An economic analysis done after the conversion showed that the robot would pay for itself in 14 months.

ONE ROBOT, TWO TASKS

Three months later, it became necessary to make a change in the basic operation. Figure 14.4 shows the body of a larger tractor that previously had been painted using a different process in another plant location. Because of increased competition and a change in marketing strategy, production of the larger tractor had to be increased, even at a loss of production of the smaller version. Much of the production work for the larger tractor easily could be sped up. However, the spray-painting process could not.

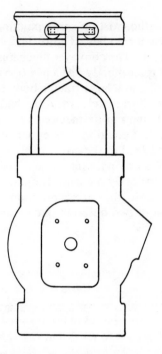

Figure 14.1. A small tractor body suspended from an overhead rack.

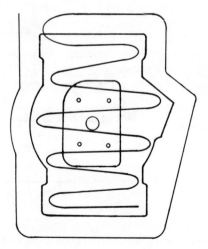

Figure 14.2. The path a spray gun will follow during a program.

It was decided that the existing spray-painting line would be used to spray both sizes of tractor bodies. Figure 14.5 shows the spray-painting path for the large tractor. The conveyer mechanism was easily retrofitted to carry both types of tractor body. At a position upstream from the spray painting booth, a small and a large tractor body alternately were hung on ice-tong-like supports. Fortunately, the robot had a dual-program feature. It was capable of retaining in its memory two distinct programs. At a simple input command, it could initiate either of them.

A limit switch was added to the conveyer line just ahead of the trigger switch that began the spray-painting cycle. The limit switch signaled if there was a large tractor body present. It did not signal if the small body was present. In either case, the correct program was run automatically, allowing the production of two different parts using the same robot.

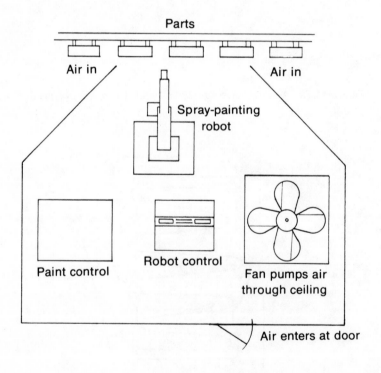

Figure 14.3.   A plan of the spray-painting booth.

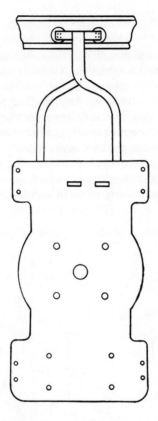

Figure 14.4.   A large tractor body on an overhead conveyer.

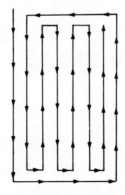

Figure 14.5.   The path the spray gun will follow during the spray program.

FOLLOW-UP

Because of the hydraulic nature of the robot, and the simplicity of both the robot and the control mechanism, virtually no malfunctions resulted in the first six months of operation. A cost analysis indicated that while the hydraulic unit would use some $800 worth of additional electric power per year, $1,472 per year would be saved in paint through more exact applications. What's more, since the robot's painting actions were the same each time, a higher quality control was achieved.

A follow-up study on workers' health indicated that those workers who had been exposed to the painting environment for up to four years were recovering from any ill effects of paint contamination.

CHAPTER 15

DEBURRING

Humongous Iron Works (HIW) makes large iron and semisteel castings under contract to other companies for use as components in their manufacturing processes.

Figure 15.1 shows a typical HIW casting loaded on a rail transportation vehicle for movement within the plant. This part has a weight of 1,147 pounds as it arrives at the deburring work station. The linear path a deburring tool must take includes all the outside edges, the interior corners of the holes and the upper protrusions' outside edges. The chamfering dimensions are shown in Figure 15.2 along with an indication that the exact position of the edge may vary up to .150 inch.

WHY A ROBOT

A robot was considered for deburring this and other types of castings because a human operator holding a powerful, high-speed deburring tool experienced great fatigue in chamfering long, irregularly shaped surfaces. Fatigue sometimes resulted in too little or too much metal being taken off, in some cases rendering the casting unacceptable by the customer. Although the chamfering process was of itself a noncritical operation, its ability to scrap out large and expensive components made it worthy of consideration for robotics.

Figure 15.3 shows another type of casting that would fall within the work load of the proposed robot.

REQUIREMENTS

In the recent past, 141 different jobs were undertaken by HIW. Mold sections and specifications for each were at hand for production on short

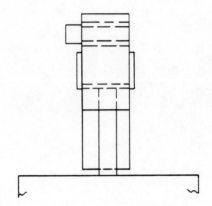

An end view of a typical casting.

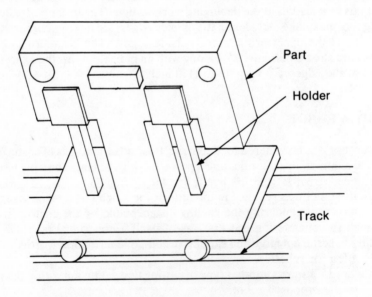

Figure 15.1. A typical casting that must be deburred.

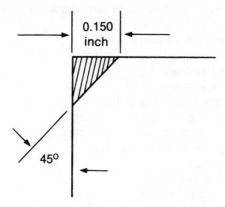

Figure 15.2.   Chamfering dimensions.

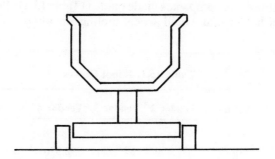

Figure 15.3.   A typical small casting.

notice. HIW determined that the proposed robot must be capable of deburring any and all of their castings.

A typical job for HIW would be the production of perhaps 50 units at a single time before a changeover was made to a different casting. Because of the short runs, fast program loading and ease of program entry were of prime concern in the selection of a robot.

Time cycles for jobs at HIW ranged from a maximum of 14 minutes to a minimum of 58 seconds; all jobs required at least some deburring. HIW engineers analyzed all 141 deburring jobs and determined that most robots capable of deburring were within the speed ranges necessary to handle each job.

HIW outlined a hierarchy of robot needs:

1. Six axes of freedom.
2. Lift 26 pounds of tool.
3. Exert uniform 50 pounds of thrust.
4. Reach all parts of the largest castings.
5. Compensate for mispositioning of part.
6. Fast enough for short time cycles.
7. Easy to program.
8. Local training available.
9. Multiple program storage.
10. Fast programming and storage of programs.
11. Delivery during the Christmas shutdown period.

After analyzing possible robots, HIW engineers determined that seven manufacturers were capable of supplying a robot that would meet company needs. Of these, six companies made quotes (Table 15.1). Prices shown include a tool holder that would accept tooling already purchased by HIW.

Table 15.1. Robot Quotes

	Vendor 1	Vendor 2	Vendor 3	Vendor 4	Vendor 5	Vendor 6
Cost Plus Tooling	$110,000	$91,860	$92,400	$92,970	$94,000	$104,371
Operating Cost	STD	STD	STD	STD	+$1,600	+$1,600
Tool Type	SP	SP	SP	FC	FC	TQ
Power	EL	EL	EL	EL	HYD	HYD
Training	+	+	-	+	+	-
Service	-	+	+	+	+	-

STD = Energy costs are better than standard
SP = Spring
FC = Force sensor
TQ = Torque sensor
EL = Electric
HYD = Hydraulic

WHICH METHOD?

Of the six companies, three offered the same basic technology to fill the needs. Vendors 1, 2 and 3 each offered a tool holder with multiple directions of spring loading. This tool holder would allow the tool to move from the expected position of the part to a position where the spring pressure was equal to an amount necessary to produce the desired depth of cut. Each vendor felt that such a tool would provide an adequate surface on the parts made at HIW.

The plant engineers, however, upon reviewing this technology, were concerned that each part, being of somewhat different material, would have a different depth of cut. To avoid this problem, it would be necessary to change the spring tension on the tool for each of the different castings. Since this would be a nonprogrammed function, considerable human attention would be required at each changeover—a requirement considered cumbersome and unnecessary.

Vendors 4 and 5 each specified a technology whereby force feedback would be monitored by sensors located in the wrist of the robot next to where the tool was joined. The robot would compensate for changes in feedback force to maintain an exact depth of cut. Figure 15.4 illustrates a circular path that would be programmed into the robot's memory. It shows how the actual position of a part could vary from the program position. For each of the small positions in space that together would create the program,

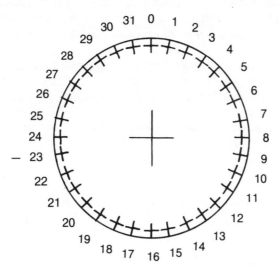

Figure 15.4. Minute changes sometimes may be necessary in the robot's path.

a positional correction vector would be established to make a response that would change the position of the robot when a force less than that required for proper deburring was encountered.

Vendor 6 specified a technology different from all the others—a torque-measuring device used in conjunction with the robot's controller to determine the depth of cut as it was being made. Although this technology was considered viable by company engineers, a price difference of over $10,000 was sufficient to reject the quote of Vendor 6.

In deciding between Vendors 4 and 5, HIW engineers considered the additional operating cost anticipated for Vendor 5's robot. The robot was hydraulic in nature and would be subject to inaccuracies caused by varying oil temperatures. It also would be subject to contamination of its working oil by dust in the plant environment and metal fragments released by the grinding wheel.

Those factors, coupled with a savings of approximately $2,600, led to the selection of Vendor 4.

MALFUNCTIONS

The robot was installed over the Christmas holidays. During the initial six months of operation, the robot had 11 malfunctions. Of these, six were malfunctions in the program. One was attributed to a bad cassette storage tape. Three malfunctions were electronic, one of these being the failure of the sensor mechanism which caused some damage to the tool. The last failure was mechanical and was the result of the tool coming loose while in operation.

No data was available regarding the percent of downtime or the monetary loss caused by robot malfunctions. In all cases, malfunctions were easily corrected and resulted in loss only until the end of a particular production run. The robot is idle between production runs, providing adequate time to repair any encountered malfunction before the beginning of the next work cycle.

CHAPTER 16

AUTO BODY GAUGING

The Profittown Auto Assembly plant, a Huge Motors Company, experienced a metal fit problem in its door operations. As a large number of components were stamped and welded together to form an assembly, minor uncontrollable variances caused rather large fluctuations in the positions of door openings and the sizes of doors.

It was possible to measure each door opening and each door and then to match a door—large, medium or small—to each opening. However, a proper measurement took as long as 70 minutes per automobile, too long for practical production.

ROBOT FEASIBILITY

In part resulting from a trade fair, at which company engineers saw a demonstration of auto body gauging using a laser camera system carried by a robot, Profittown Auto Assembly made a request for a quote to several

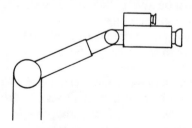

Figure 16.1. A camera and laser light source at the end of a robot.

robot companies. Several vendors involved in large-scale measurement of car bodies were able to quote equipment such as that shown in Figure 16.1.

Each of their systems operated similarly. A laser light source was directly coupled to a camera that could interpret both the angular displacement and intensity of the laser beam. A robot would carry the camera unit to the item being measured, say a door opening (Figure 16.2). Any deviation from the desired position would be registered by the camera and indicated electronically to some outside computer equipment. The accuracy of the robot itself in positioning the camera would be critical to this application in that a robot deviation would not be witnessed by the camera and therefore would cause a misinterpretation of the camera's data.

Because of the large distance between the points to be measured, it was determined that two robots would be used to measure the cars.

CRITERIA

The robots would need tracking ability so they could be unsynchronized with the moving cars. The other robot criteria were as follows:

1. Robot vision system deviation in each position of the work cycle must be less than .012 inch.
2. The robot must be able to settle in each of the 14 inspection stations within 54 seconds. (Settling time is the time necessary for the robot to remain stationary so that any wiggle or jar of the camera will have dissipated before a measurement is taken.)
3. The robot must be able to move along with the car and to stop if the line stops, avoiding damage to the robot equipment.
4. The robot must have a relatively long reach—it must be able to reach all of the points on one-half of the car body from its stationary position.
5. The robot must be easy to program and update.
6. The robot must be able to be coupled with the computer selecting the doors for door openings.
7. The camera must be modular so that should some damage occur, the camera could be replaced quickly with a spare.

Two vendors were able to meet all criteria. One was selected based, not strictly on price, but on its past history of successful dealings with Huge Motors Company. Figure 16.3 shows the completed installation—two robots on either side of the moving conveyer line for 100 percent inspection of car bodies.

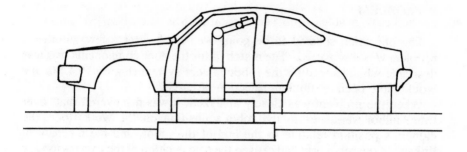

Figure 16.2. Robot inspection of car body.

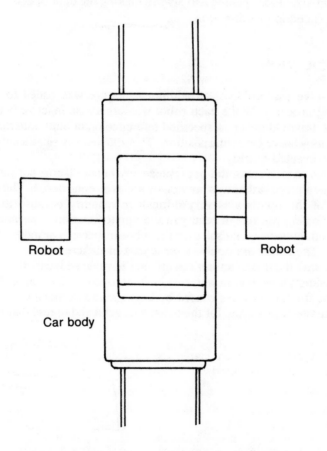

Robot

Robot

Car body

Figure 16.3. Two robots inspect a car body as it passes by on a conveyer.

A PROBLEM

The operation seemed to be going well until, after a few months, a problem was discovered. The match fit for the door components was less desirable when gauged by the robot camera system (Figure 16.4.) than it would have been by random selection of parts.

When the problem was closely analyzed, it was discovered that, over time, minor variances in the robot's position on the work floor, the camera's position relative to the end of the robot, and the mechanical linkage of the robot arm had caused the true position of the camera to vary while in use. The camera had indicated a bias of position for all of the parts, for example, sometimes registering a too-small door opening as a medium-size door opening and always making the door opening appear to be displaced in one direction.

THE SOLUTION

To solve the problem, an additional feature was added to the robot gauging process. While each robot was otherwise inactive between car bodies, it would turn to a specified reference point built onto its base and check its relative mounting position. This allowed any significant deviation to be corrected quickly.

In this case, a robot did not replace any human labor, but rather made possible a process that was previously not even considered. The economic value of the robots was very difficult to quantify because the product improvement was one of beauty in addition to function. The manufacturer reported, however, that sales for the robot-inspected car model were quite brisk. The robots no doubt deserve part of the credit.

One way the manufacturer quantified the value added by the robots was to consider the savings in warranty claims based on the misalliance of doors and the fracture of glass caused by improper door-opening size. On this basis alone, a buy-back for the robot was achieved in less that two years.

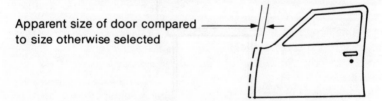

Apparent size of door compared
to size otherwise selected

Figure 16.4. A correction can be required in the size
of door needed when robot inspection is used.

CHAPTER 17

TAPE RECORDER ASSEMBLY

The Fun Time Electronics Company makes, among other electronic equipment, an inexpensive tape recorder for use with home computers and video games.

The final assembly of the tape recorder was a bottleneck and was the most costly single operation in the manufacturing cycle. In this operation, two workers assembled various components in a plastic base. The time cycle was 92 seconds. An analysis showed that a tape recorder could be made every 38 seconds, should the final assembly be sped up. An engineering study was undertaken on the feasibility of increasing productivity by using robots for assembly.

ASSEMBLY OPERATION

The tape recorder final assembly involves the parts shown in Figure 17.1. The base is held in position. To it are added a circuit board, a layer of insulation material, an electric motor and drive unit, and a cover containing the push buttons for operation.

The insulation material comes in sheets and is set on a pallet in front of the worker. Because there are 2,400 sheets on a pallet, many production cycles may run from a single pallet.

The modular design of the unit allows it to be assembled without any soldering or direct wiring of components. Additionally, only two screws are necessary to hold the motor and circuit board in place.

Parts are delivered to the final assembly work station in lightweight fiber boxes. This system of containerized parts delivery is used throughout the manufacturing plant and allows parts to be delivered from one station to another without a continuous and synchronized assembly line. The base is

123

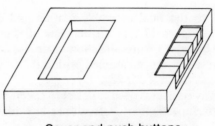

Cover and push buttons

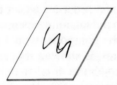

Electric motor and drive unit

Insulation material

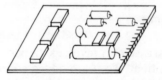

Circuit board

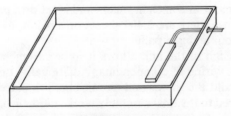

Base/connector

Figure 17.1.   A view of parts to be included in assembly.

fabricated adjacent to the final assembly station and does not need a shipping container. Figure 17.2 is a side view of a container used to transport the circuit boards. Figure 17.3 shows a similar container and how the motor and drive units are positioned within it. Several motors are located in exact positions in a dunnage layer.

Dunnage by the way, is a term used to denote a specific container holder that protects parts during shipment. Oftentimes, the dunnage is a molded plastic sheet that fits parts snugly and holds them in relation to each other and within a shipping container. The term dunnage is also used to denote the layers of protective material placed between layers of parts shipped in a single container. Figure 17.4 shows the shipping container for the plastic covers of the tape recorder.

ROBOT FEASIBILITY

In designing the robot assembly operation, a light source and light sensor were planned—one on each side of the conveyer (Figure 17.5). When an individual base component reached its proper position, the light beam would be broken and the conveyer motor would be stopped abruptly.

An early problem encountered by the engineering team evaluating robot use was the different shapes and sizes of parts to be handled. Each of the components had a different dimensional structure making it difficult for all components to be grasped with the same gripping system. Additionally, a time and work analysis indicated little possibility of a single robot being able to perform all of the tasks within the allotted time frame of 38 seconds. So, the job was broken down into three processes: Pick up components, place them into position, and fasten the unit together using screws.

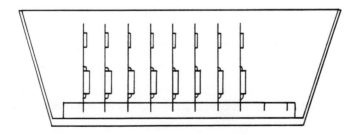

Figure 17.2. A tray is used to transport circuit boards.

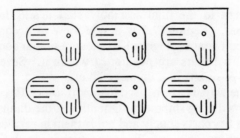

Figure 17.3. Motors are in a low-sided box.

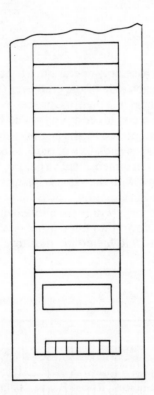

Figure 17.4. Plastic covers are stacked, one upon another.

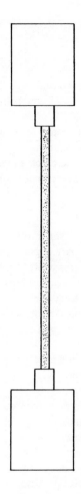

Figure 17.5.   A beam of light is used to position the main subassembly.

Three robots were planned to handle these processes. Figure 17.6 shows the working area floor plan. Robot A was assigned to handle the circuit boards and motor components. Robot B was assigned the insulation layer and cover, and Robot C, the fastening of the units with screws. Robots would alternate steps for the most efficient use of time. This also would allow a robot, while waiting for its counterparts to perform tasks, to remove empty shipping containers and place them in an out-of-the-way location.

Table 17.1. Tape Recorder Assembly

Prestart Conditions

1. Robot A grasps circuit board
2. Gripper indicates correct circuit board position
3. Robot A lifts circuit board
4. Robot A carries circuit board to preassembly position
5. Robot B lifts insulation material
6. Limit pressure switch indicates material in position
7. Robot B moves to preplacement position

Robot Program

1. Base moves along conveyer into position
2. Beam break indicates part in position and stops conveyer
3. Conveyer stops, indicates to Robot A ready for assembly
4. Robot A places circuit board in position
5. Robot B moves close to conveyer
6. Robot A releases circuit board
7. Robot A withdraws from base
8. Robot A cycles gripper
9. Gripper indicates part released
10. Robot B places insulation material
11. Robot A grips motor unit
12. Robot A gripper indicates part in position
13. Robot B releases insulation material
14. Robot A lifts motor unit
15. Robot B withdraws over to cover
16. Robot A carries motor into position
17. Robot B indicates to Robot A position clear
18. Robot A places motor in position
19. Robot B grasps cover
20. Robot A releases motor
21. Robot B lifts cover
22. Robot B receives signal from gripper part in position
23. Robot A receives signal from gripper part released
24. Robot A withdraws
25. Robot B receives signal from Robot A base clear
26. Robot B moves cover in position
27. Robot C moves arm under cover ready to screw
28. Robot C applies fasteners
29. Signal from fastening unit screw is tight

Table 17.1. Continued

30. Robot C disengages screwdriver
31. Robot C withdraws
32. Robot B positions cover
33. Robot A grips circuit board
34. Robot B withdraws from conveyer
35. Robot B signals conveyer release, finished tape recorder
36. Robot B grips insulating layer
37. Repeat program

A sample program to facilitate the robotic tape recorder assembly is shown in Table 17.1.   Although the program is oversimplified, it demonstrates the basic operations necessary for the complete assembly of the tape recorder unit. In reality, hundreds of programming lines would be necessary for each robot to adequately communicate with each other and surrounding equipment.

SPECIAL EQUIPMENT

In planning the robotic conversion, it was necessary for gripping units to be designed specifically for the job at hand. To facilitate the positioning and gripping of the circuit boards by Robot A, a gripper was designed with a 90° hook on its end. This allowed the robot to pick up a circuit board–which was in proximity to others—and place it in the tape recorder base without touching the base. This gripper could also pick up the motor drive unit (Figure 17.7). A plastic coating would be applied to the contact surfaces to avoid damaging the components as they were carried. Sensors would be placed within the gripper that would signal both a presence of part and a correct part position within the gripper. This would eliminate false assembly.

Robot B's gripping unit—designed to pick up the cover and insulating material—was far simpler (Figure 17.8). Vacuum cups would be used to sense the presence of material and lift it. Robot C's job would be to fasten the assembled unit with screws. It would be fitted with a metal plate holding two standard automatic screw guns (Figure 17.9). The self-feeding fastening screws would be supplied to the guns on a paper roll.

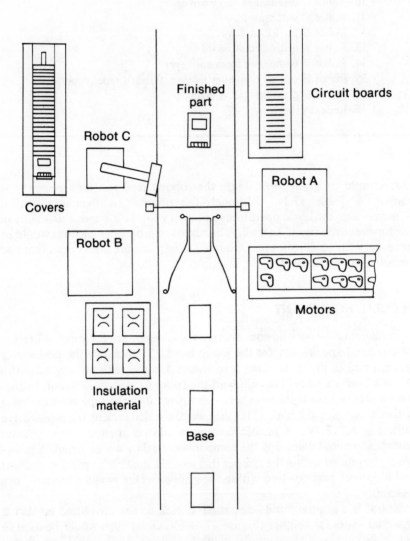

Figure 17.6.   A plan of the three-robot assembly station.

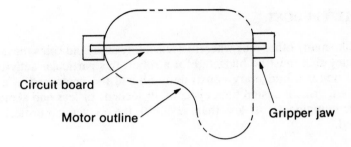

Figure 17.7. Proper design of Robot A's gripper
jaws allows the holding of several parts.

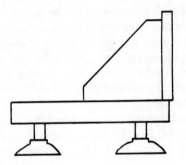

Figure 17.8. Robot B's gripper for covers and insulation.

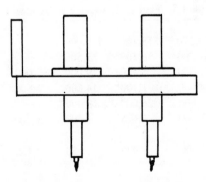

Figure 17.9. A dual screw-driving head for Robot C.

LIMITED BUDGET

Unlike many other jobs wherein a feasibility study indicates the amount of money that must be budgeted for a robot, this particular activity was placed under a budgetary constraint. Should the engineers be able to achieve a critical design time cycle of 40 seconds or less and should the robotic equipment cost less than $250,000, the conversion project could proceed.

VENDOR QUOTES

Table 17.2 shows the proposed equipment and quotes by robot and subsupplier vendors. The request for quote from Fun Time Electronics specifically stated that all sub-items must be called out separately and that it was the intention of the company to purchase each from the vendor with the lowest price. If a vendor wished, however, it could also quote the entire system, taking the responsibility of system integration. In Table 17.2, the figures with asterisks are the lowest for each piece of equipment. The total price, if each component were to be purchased from the lowest individual bidder, would be $140,560.

Fun Time estimated that the installation of this equipment would require an additional 70 hours of effort by electricians. It estimated that six 40-hour weeks would be spent by each of two engineers preparing for the

Table 17.2. Robot and Component Quotes

Item	Vendor 1	Vendor 2	Vendor 3	Vendor 4	Vendor 5	Vendor 6
Robot A	$64,270	$58,160*	$61,000	—	—	—
Robot B	61,200	58,160	55,500*	—	—	—
Gripper A	24,000	6,500*	19,000	$9,000	$12,000	$22,000
Gripper B	4,000	1,100*	14,000	1,800	6,000	12,000
Robot C	29,000	26,000	18,000*	—	—	—
Auto Screw Unit	3,800	1,675	4,400	1,800	4,000	900*
Light Guide	1,700	900	650	450	1,000	400*

* Lowest priced

installation and that one or the other would have to be present also during the 70 hours of installation. This amounted to 550 engineer hours. Wages and benefits were $24 per hour for electricians and $30 per hour for engineers, resulting in a total of $18,180 in wages for installation. This would bring the total estimated cost for the installed equipment to $158,740.

However, another factor had to be considered. Vendor 2 had indicated that if it received the order for all of the components, it would supply equipment, installation and programming for $150,000. Fun Time estimated that if this option was selected, the only company manpower needed would be one engineer to act as project manager for two weeks. That cost would be $2,400 (80 hours x $30). So, the total installed price would be $152,400.

Fun Time awarded Vendor 2 the complete job. Although the money saved by this selection was not great, the company felt there was a greater likelihood of the job being completed on time and within budget by working with one vendor.

MALFUNCTIONS

The decision was a sound one. Installation of the equipment proceeded according to schedule. But once on the factory floor, the equipment did not function as planned. There was a high rejection rate of tape recorders because of the mispositioning of the insulation layer and the fracturing of the plastic cover during assembly. Many weeks of effort were required by Vendor 2 engineers and service personnel to solve the problems. The solutions involved using insulation material with a slightly tacky bottom surface, to better hold in position, and reducing the power on the screw gun.

In retrospect, it was determined by management that there had been no vendor error. Rather, the original design of the equipment did not take into account variances in the work cycle. Had the company used its own personnel for the installation, it is doubtful whether their expertise would have been sufficient to solve the problems in a timely manner.

According to common practice, if a complete set of robots are to be installed, the installation is done at an off-site location. There, the system can be completely tested without the necessity of shutting down an assembly line. However, inexperience led the electronics company to believe the installation could be done during a plant shutdown and all of the variables could be brought under control before the scheduled start-up.

FOLLOW-UP

An analysis of the robotic assembly operation indicated a high percentage of uptime. The original tape recorder assemblers, assigned to other jobs, occasionally watched their old jobs being done by robots. One of the workers remarked, "While the assembly job was not that bad, I'm glad the company didn't try to speed it up this fast while I was doing it."

CHAPTER 18

SPOT WELDING

United States Motors produces automobiles in many plants throughout the United States.

In operation 62 at one of their plants, the back seat support panel and deck lid support panel are spot welded together. In the same operation, the deck lid hinge support bracket is also welded to the deck lid support panel. Figure 18.1 shows a car body and indicates by x's the positions of welds that must be performed.

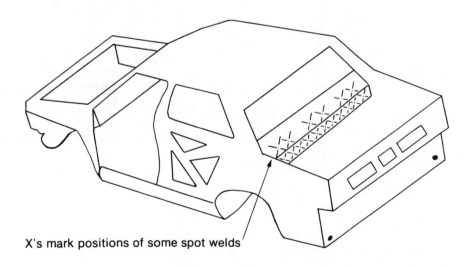

X's mark positions of some spot welds

Figure 18.1. A car body showing the position of welds.

THE PROBLEMS

The company felt the welding operation might be a good job for a robot. First, the cost of the part being produced was great, requiring accurate, high-quality work. Second, the work was cumbersome for a human worker who had to climb aboard the automobile, manipulate a heavy, spark-producing, potentially dangerous piece of equipment, work in a hot environment, hurriedly perform his task, and then leave the automobile before it was allowed to move to the next work station.

To assist the human operators in the existing plant, an overhead lifting mechanism for the welding gun was used (Figure 18.2). A counterweight closely equalling the total weight of the welder, gun and cable system was placed beyond a series of pulleys and joined to the welding gun by a cable. This allowed the human operator to operate the gun with less fatigue than if he had to lift the full weight of the equipment. Figure 18.3 shows how the

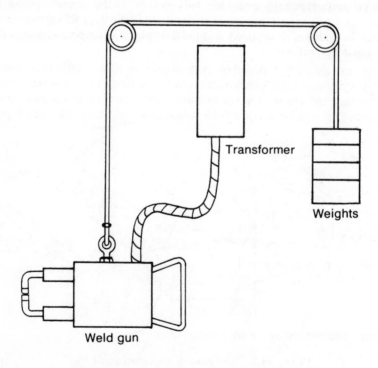

Figure 18.2. A manual weld gun is suspended in air by weights and pulleys.

jaws of the welding gun clamped together the pieces of metal to be welded. Then, the gun would supply great current levels to melt the metal between the electrodes, causing the pieces to weld together. In operation 62, there were 22 welds to be done during the 44 seconds the automobile was at the work station.

Particular problems with this manual operation encountered by management were: (1) missing welds resulting in car rattles and misalignment of components, (2) welds in the wrong position causing misalignment of body parts or bad finish on the body, and (3) high worker absenteeism and requests for relief during the normal work period.

It was estimated by company engineers rather familiar with robot equipment that one robot could replace the one human worker at operation 62 and that the worker could be reassigned other less physically demanding welding tasks. For the human worker, a $14.42 hourly rate, plus an estimated additional 50 percent for benefits, brought the total hourly rate to $21.63— $173.04 for an eight-hour day.

THREE POSSIBILITIES

The engineers determined that there were three possibilities to automate operation 62.

One was a gantry-type robot system (Figure 18.4). With this system, beams are erected through which the work passes. A robot is placed on a traversing mechanism atop the beams. Often, this traversing mechanism resembles an overhead crane with a rigid robot attached to it. The type of

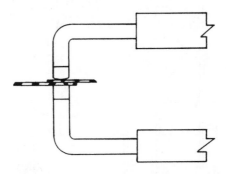

Figure 18.3. The jaws of a welding gun both
clamp the metal and conduct electricity through it.

gantry system envisioned for this operation would have a transfer mechanism indexing the length and width of the automobile from high above, an arm mechanism that would linearly position downward toward the car and several other axes to manipulate the gun to correct angles.

Figure 18.5 shows the second robotic possibility. Still based on the gantry principle, this alternative employs a more conventional-type robot, mounted upside down. As with the first type, the robot is carried aloft on a series of beams and can move along the length and width of the automobile. An articulated-type arm, like that of a regular floor-mounted robot, moves the welding gun into correct position and administers the weld.

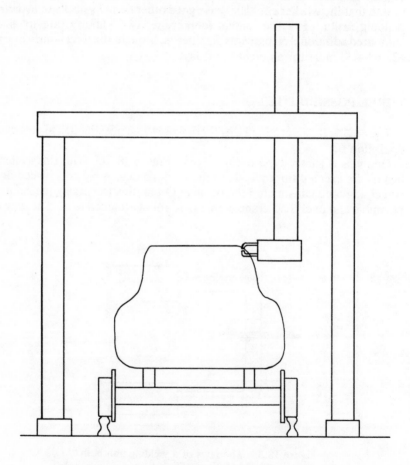

Figure 18.4. A gantry-type robot welding a car body.

Figure 18.6 shows the third possibility—a more conventional approach to robot spot welding. A spot-welding robot is positioned on each side of the car. (An extreme reach would be necessary for one robot to do all the welds from one side of the car.) The robots carry guns into position, accomplish their task and move to the side of the car, allowing it to pass.

COSTS

Several vendors were asked to quote robot systems and responded with the prices shown in Table 18.1. Vendor 1 proposed the first gantry-type system at a cost of $142,729.

Vendor 2 proposed the gantry-type system with an inverted conventional-type robot. The robot cost was $115,400; cost of a custom-made track system was $22,000.

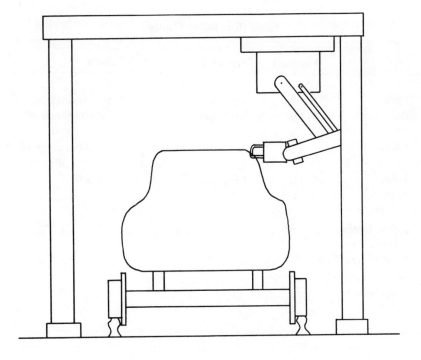

Figure 18.5. A traditional, articulated-arm may be mounted upside-down.

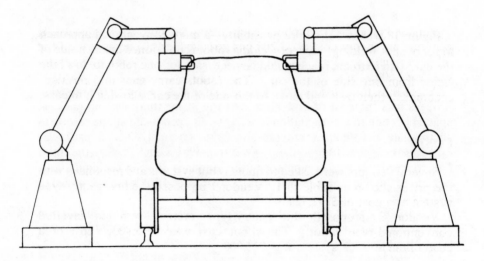

Figure 18.6.   Floor-mounted robots must work from either side.

Table 18.1. Robot Quotes

	Vendor 1	Vendor 2	Vendor 3
Type	Gantry	Inverted	Conventional
Base Price	$142,729	$115,400	$ 88,600 (each of two)
Extra		$ 22,000 (support mechanism)	$ 3,200 energy (each of two)
Total	$142,729	$137,400	$183,600
Pay-back— 3 shifts	1.1 years	1.1 years	1.4 years
Robot Uptime	99	98	97
Ease of Repair	+	-	+
Ease of Installation	+	+	-
Auto Program Adjustment	+	+	+

Vendor 3 proposed two robots, each costing $88,600. Each robot would generate an energy cost of $3,200 per year above the energy cost of the other proposed systems. So, the total first-year cost of Vendor 3's system was $183,600.

Another factor considered by United States Motors during the selection process was the cost of downtime. At the plant, there was no physical space between the work stations for car bodies produced at one station to accumulate before they entered the next station. When a particular operation ceased to function, only a few moments went by before the entire line stopped. It was computed by the company that on the continuous portion of the line that included operation 62, $872.41 of value was added to the vehicle.

If a robot system were installed at operation 62 and if a malfunction occurred, the line need not stop, providing that the robot could be moved out of the way. At a subsequent work station, specifically operation 65, human welders could perform welds missed at operation 62. The cost for this extra work at operation 65 was calculated at $35 per hour. It was estimated that one-sixth of the cars would need work done at operation 65 (for failure not only at operation 62 but also at other stations) at a cost of $24 per automobile. One hour of lost production equals a loss of $35 plus an estimated $480 in repair costs.

To calculate the cost to United States Motors for the expected downtime, the engineers considered an overall cost of $5.15 per hour times an available 6,000 hours, yielding, at 1 percent of downtime, $30,900. Each percent of downtime yielded by the robot would have this negative value on the automotive production.

SELECTION

A selection of Vendor 1's gantry robot by United States Motors resulted in increased production capability. During the vendor's warranty period, the robot ran with an average 96.7 percent uptime.

A major factor in the unexpected lower numbers was the reluctance of plant personnel, after a robot repair had been made, to restart the robot. Since the robot's work could be done at operation 65, the repaired robot was often left idle until a lunch break or shift change would allow a few test cycles without risk that the robot would completely stop the line. No correction was made to this situation as far as we can determine.

CHAPTER 19

LOADING AND UNLOADING BOBBINS

East Carolina Mills is a producer of raw filament used in weaving and textile processes. By chemical mixture, a rayon material was produced in large vat quantities and was forced under pressure through fine openings in a metal plate. This produced very thin filaments of rayon fiber at such rates that miles of it extruded from each orifice every minute.

Because of the processes involved, it was necessary that the filament continuously be drawn away from the orifice. This was accomplished by a vacuum draw-tube that drew the freshly formed filament from the metal plate on the building's third floor to a lower-level room. An opening in the pneumatic draw-tube allowed a worker to catch a strand of fiber with a hook and begin winding the fiber on a cone-shaped bobbin on a spindle (Figure 19.1). The hook used by the operator had a sharp edge in its crook (Figure 19.2). The operator grasped the moving filament, twisted it several times about the bobbin, cut off the loose end, and by means of a foot pedal began a very high-speed rotation of the spindle.

When a bobbin was completely wound with fiber, a weight scale triggered a break on the spindle, and it came to a stop. At that time, since there was no pull against the fiber by the spindle, the pneumatic draw-tube had sufficient suction to again pull the fiber down. The operator now used the hook to cut the fiber, removed the finished bobbin of fiber (Figure 19.3) and placed it in a shipping container.

THE PROBLEMS

There were a couple of problems with this process. One was that during the time that the operator transferred the completed bobbin—before a new filament was stranded onto an empty bobbin—the pneumatic draw-tube continuously took the newly formed filament and dumped it into a

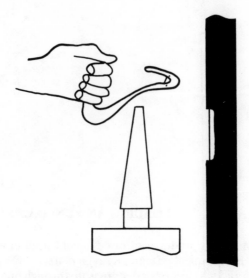

Figure 19.1.   A pneumatic draw tube and the hook
used to retrieve a strand of filament from it.

basement scrap container. Because of the process, there was no way to
preserve this otherwise useful and perfect thread. The waste filament was
used for other purposes, specifically seat-cushion packing. However, the
value of the material as packing was considerably less than that as
pristinely wrapped bobbins of filament for textile production.

An additional problem was filament tangling during shipment. Although
bobbins weighed 31 pounds each, the material on them was extremely
fragile. The filament has a high tensile strength; yet, if the bobbin was
bumped or improperly grasped, individual strands of fiber rolled one on top
of another. Such a rolling effect—during high-speed weaving—caused a
tangle in the line. This either caused the scrap of finished cloth or at least
stopped the machine so the tangle could be cleared. For this reason, the
plant maintained a high quality control check. And so did customers.

Figure 19.4 shows the manual picker used to pick off a full bobbin from a
spindle and place it on a specially prepared shipping rack. Figure 19.5
shows the element within the shipping rack that received an individual
bobbin. A clearance slot in the side of the bobbin carrier allowed the picker
to slide into and out of position easily.

Because of the relatively high value of the rayon fiber and its percentage
of loss, an analysis was undertaken to consider more cost efficient
production methods. A time and motion study produced the following
results: A bolt of rayon filament required between 93 and 96 seconds to

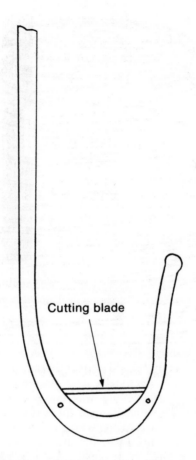

Figure 19.2.   A closeup of the hook end.

spin. Depending upon minor variations in the density and diameter of fiber, the weight was subject to change. On the average, 42 seconds were required to change a bobbin, store it, place a new empty bobbin on the spindle, retrieve and wind a fiber element and begin the next spinning operation.

This exchange time accounted for a rayon filament loss of almost one-third. A value was placed, based on the cost of the filament production machinery and the cost of the raw material chemicals, at $.326 per second of production.

Figure 19.3.   A bolt of filament.

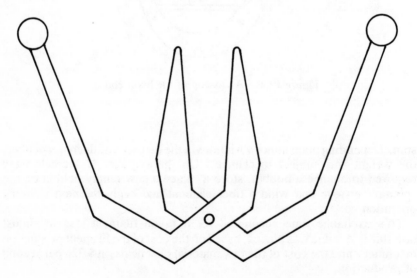

Figure 19.4.   A manual bolt lifter.

Figure 19.5.   A portion of the transport rack that holds the bobbin.

It was also discovered that seven percent of the completed bobbins were scrapped by either the East Carolina Mills quality control department or customer receiving inspection departments. The study recommended that a method be found to shorten the exchange time for the bobbins and, further, that an automation process involving robots be considered.

ROBOT FEASIBILITY

ABC Automation, Inc. was contracted to conduct a feasibility study on the use of robots in rayon filament production. Table 19.1 shows a cost analysis based on manual production of the filament and standard scrap rates.

A total cost per bobbin of rayon, up through the spinning and packing point, amounted to $46.174. The selling price was $50, yielding a return to the mill of approximately $3.83 per bobbin. This gross profit was deceiving, however, because of the marketing and selling costs, the transportation and ownership costs, and the occasional loss due to industry fluctuations. The gross profit margin of approximately seven percent had been considered adequate until the efficiency study began.

As a part of the feasibility study, a rough design was made by ABC for two types of robot gripper. Gripper 1 (Figure 19.6) allowed for the automatic pickup of bobbins once they were spun with filament. Gripper 2 (Figure 19.7) had a dual nozzle vacuum tube allowing both the capture of filament moving through the vacuum draw-tube and the suction lift of empty bobbins from a bin.

Thinking that with these gripping units and standard robots it would be possible to perform all of the manual operations, ABC produced a computerized analysis of the robot moves. Table 19.2 shows an analysis of all the times necessary for completing the most critical portion of the program. (The period during which filament is being dumped by the pneumatic draw-tube is the most critical.) Total time was 24.5 seconds.

Table 19.3 shows an estimate by ABC of the potential increased earnings of East Carolina Mills should a robot be used. It is based on a zero-scrap expectation and on a full 24 hours of production for 360 days.

ABC's estimate for the cost of equipment necessary to automate this job is shown in Table 19.4. No equipment was actually selected or quoted, rather a budgetary estimate was arrived at based on ABC's experience in other jobs. At a potential increased revenue of $6,111 per day, $177,000 of robot equipment could be paid for in about 29 days of production. After the

Table 19.1. Manual Method

Labor Cost	$5.62 per hour
Benefits Cost	$2.21 per hour
Afternoon-Shift Premium	$.15 per hour
Night-Shift Premium	$.25 per hour
Average Production Per Day	610 units
Average Scrap Per Day	42 units
Total Daily Labor Cost	$191.12
Total Filament Cost	

$$\frac{86{,}400 \text{ seconds} \times \$.326 = \$28{,}166}{610} = \$46.174 \text{ per bobbin}$$

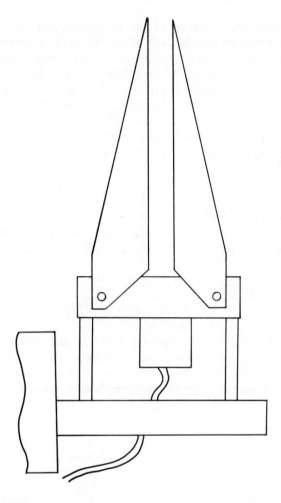

Figure 19.6.   A gripper to automatically lift bobbin.

robot itself was paid for, assuming for the moment that there would be no maintenance costs, only $28,166.40 would have to be expended to achieve an on-site revenue of $36,450. This would amount to an approximate 29 percent instantaneous return on investment, an unbelievable increase over the seven percent return heretofore experienced by East Carolina Mills.

At this point, it might be expected by everyone, as was the case at East Carolina Mills, that there would be no stopping the use of robots.

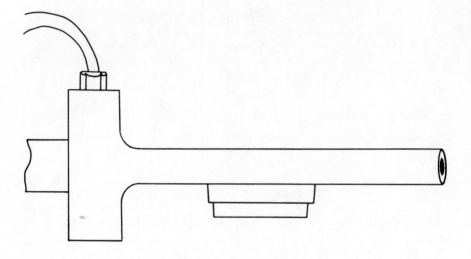

Figure 19.7.   A pneumatic gripper to lift empty bobbins and capture filament.

Table 19.2. Time Analysis

1 robot to unload, 1 robot to wrap filament
3 sec. to stop spindle
4 sec. to orient spindle for gripper
6 sec. to clear
1 sec. for bobbin gripper to enter working area
2 sec. to place bobbin
3 sec. to travel to filament and locate
1 sec. pause for vacuum capture
1 sec. to approach bobbin
2.5 sec. to wrap and cut
_____
24.5 sec.

Table 19.3. Robot Method

Zero scrap, full 24 hours

$\dfrac{86,400 \text{ sec. per day}}{118.5 \text{ sec.}}$ = 729 units per day

$50 per unit = $36,450 value per day
                -30,500 cost per day
                _____
                $ 5,950 extra revenue per day
                x 360 days
                _____
          $2,142,000 per year

                $5,950 extra revenue per day
                + 191 saving in labor costs per day
                _____
                $6,141 total

Pay-back period = 29 days (177,000 ÷ 6,141)

Table 19.4. Robot Costs

Bobbin Picker Robot	$80,000
Bobbin Gripper	10,000
Bobbin Placement Robot	60,000
Suction Gripper	7,000
ABC's Project Management Fee	20,000
	$177,000

Figure 19.8.   Dual-spindle winding apparatus.

A BETTER WAY?

However, the money paid ABC for a feasibility study was perhaps the best $4,000 East Carolina management had ever spent. Rather than merely accomplish their base task of determining if robotics were feasible and, if so, selecting equipment, ABC considered other manufacturing alternatives.

ABC discovered that a human worker was actually faster at some portions of the filament-winding operation than was the robot. Specifically, most of the human workers were able to quickly capture the filament, wind it three times about the empty bobbin, and start the bobbin to spin. A human worker fared less well on the entire operation, however, because of the time necessary to load the picking device with a bobbin and gently place the bobbin on a transfer rack. Because of the perceived waste during the loading cycle, the workers caused much of their seven percent waste in finished product by too hurriedly placing the finished bobbin on the rack.

Noting this, ABC considered the feasibility of placing two workers at the same work station doing much the same function as the two robots, thereby reducing filament scrap. However, because of safety conditions and a very cramped working space, it was not possible for two humans to work on the same filament-winding operation.

ABC conceived the best method to accomplish the operation. It involved a dual-spindle winding apparatus (Figure 19.8). Since it took between 93 and 96 seconds to wind a bobbin, a human worker could, with great efficiency, place a completed bobbin on the rack during the work cycle of another bobbin. The first operator could spin one bobbin full, cut the filament, cause the slide mechanism to index, within four or five seconds capture the filament once again, and begin the spinning of the second bobbin. While this second bobbin was spinning, the second operator could leisurely remove the full bobbin and place an empty bobbin on the spindle. Table 19.5 shows the economic factors involved in using the

Table 19.5. Dual-Spindle Winding Method

Total time delay during scrap cycle—11 seconds
Production per day—815 units

Value per day	$40,750
Cost per day	30,500
Savings per day	$10,250
	x 360 days (1 year)
	$3,690,000

Pay-back period = 6 days ($61,000 ÷ $10,250)

dual-spindle technique. Total cost for dual-spindle retrofit was $61,000.

Although in this particular case it was possible to automate a job and remove a human worker, thereby saving the company money, this was not the best way to cut costs and increase profit. A combination of human and machine labor, at far less cost than robotic equipment, turned out to be the best possible method.

An additional benefit was the job satisfaction of the human workers. Scrap rate was down substantially and production was increased, but a worker had a less hectic pace to maintain. A job-satisfaction questionnaire was filled out by workers after they had used the new equipment for 30 days, and 82 percent indicated they were more satisfied with their work life than they had been under the old system.

CHAPTER 20

STACKING AND LOADING CASES

The Koala Kola Company is a producer of beverage syrup and bottler of three types of carbonated soft drink. The most popular brand name is Koala Kola—some 6,000 12-ounce cans are produced per day. The Koala Kola bottling plant had been in production for a number of years and used some processes considered outdated by modern standards.

Figure 20.1 shows operation 27, wherein filled and sealed cans were transported along the roller conveyer from a seal test operation to the load and stack operation. The conveyer, which split into two parallel lines, allowed the accumulation of cans at the pickup station and their placement

Figure 20.1. Koala Kola can on conveyer.

in six-pack cardboard containers. As each group of six cans fell into place, automated machinery sealed the end of the cardboard container.

Koala Kola engineers wished to examine the feasibility of automating the stacking of the six-packs into cases for loading onto delivery trucks. In the filling process, one can was filled every three seconds, necessitating the removal of a six-pack approximately every 18 seconds.

A time-and-motion study indicated that when each case—containing eight six-packs—was filled, there was some delay in the removal of the next six-pack while the filled case was loaded onto a pallet for eventual removal to a trucking bay. When too many cans accumulated at the work station, a filling nozzle stopped its work, and production of the beverage was temporarily halted.

ROBOT FEASIBILITY

Because engineers at Koala Kola were completely unfamiliar with robotics, they enlisted the aid of a robot vendor to do an initial study of the operation and, if it seemed feasible, quote robotic equipment to replace human labor.

The robot vendor happened to be a supplier of other automation-type machinery, and it was clear to the vendor that an already existing piece of dedicated machinery was available to do the job. The machine could wrap beverage containers into six-packs, place the packs in standard cases, seal the cases and, by use of accumulating conveyer, load the cases onto pallets.

In this instance, the cost of even an inexpensive robot to replace human labor had a greater cost than a single-purpose machine that could do all of the operations, including that done by the existing cardboard-wrapping machine.

Koala Kola does not have a robot in its bottling plant. While at some point in the future the company might have a true need for one or more, its current needs are best met with single-purpose machinery. The best solution for increased productivity in this particular case was not to use a robot at all.

CHAPTER 21

INSPECTION AND DEBURRING

Central Electric Supply produces, among other products, a domestic watt-hour meter for sale to local electric companies.

The main housing of the meter is produced from an aluminum casting weighing approximately one pound. Because of the intricacy of the casting (Figure 21.1) and the necessary accurate fit of the other components, each part must be inspected and burrs removed where they occur on mating surfaces.

The parts are produced at a rate of 11.5 seconds each on a continuously moving conveyer line. The last work station before assembly is the manual inspect and deburr station. Because there are many fluctuations in the part itself, it is common for a large number of parts to be produced virtually burrless. At these times, the operator does mostly inspection and very little mechanical work. At other times, it is difficult for the operator to keep up with the deburring on each part, and production lags at this work station.

ROBOT FEASIBILITY

A study was done by the Central Electric Supply's engineers to determine the feasibility of automating the inspection and deburring process. After consultation with several robot companies, Central Electric determined that a single robot—either grasping the part and moving it by a deburring tool or carrying a deburring tool to the part—would be able to do the job. However, a longer time cycle would be required than the manual process.

An analysis by one vendor, given the length of deburring path and complexity of motion, indicated that at least one minute would be needed to continuously deburr all the critical surfaces. In this proposed robotic

Figure 21.1.   Aluminum casting of a watt-hour meter.

installation, all parts would pass by the deburring tool—whether or not they needed deburring—as the robot had no way to determine the need to deburr. It would be possible to get the job done in approximately the same time as required by the manual method by using multiple robots. Figure 21.2 shows how a conveyer might be arranged to move the parts to six different lines, each receiving the number of parts that could adequately be handled by a single robot.

It was estimated that a robot capable of doing the job would cost $60,000 and a combination of fixturing and gripping equipment would amount to $15,000, for a cost per robot of $75,000. Since six robots would be required, total robot cost would be $450,000. Conveyer equipment to segment parts into six groups was estimated at $22,000, for a total of $472,000 for the automation process.

An alternative was proposed by a different vendor. In this operation, a vision system would inspect the part for burrs and then command the robot to move only those parts needing attention to the deburring tool. This approach would be very similar to that used by the human operator—examining the part for burrs and then performing only the deburring needed.

Figure 21.3 shows different types of burrs—each of which would be interpreted differently by a single fixed-position vision system. The camera's line of sight might cause it to miss some burrs that would

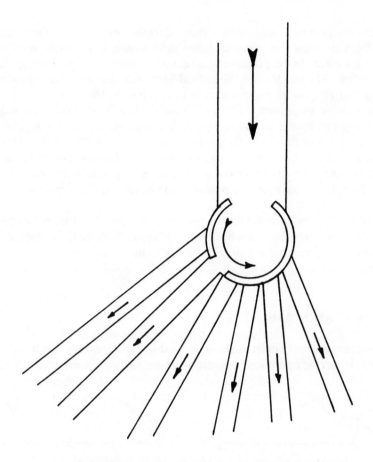

Figure 21.2.   A multiple-path conveyer system.

nonetheless be damaging to the finished watt-hour meter assembly. For this reason, it would be necessary to mount the vision system on a robot that could move it to various positions so a three-dimensional picture of the part could be developed in computer memory.

It was estimated that with this type of system, 10 seconds would be required to scan all of the features of a part, the average robot deburr cycle time would be 10 seconds, and five seconds would be needed to pick up and put down the part. This would average 25 seconds per robot cycle.

It was determined that only three robot vision systems would be needed in this particular operation. A word of caution came from the vision vendor, however, indicating that certain part anomalies and misinformation on the computer line and vision link would produce significant errors in the positioning direction given by the vision computer to the robot. The result would be critical damage to the part by overburring or underburring, at an estimated rate of one part in 50.

The vendor estimated costs as follows: The basic vision package including its computer would be $110,000, a robot would be $60,000 and a proposed gripper unit, $15,000, for a system cost of $185,000. Three systems would be needed for a cost of $555,000. However, on this job, it would be necessary to break the main transfer line into three parts, and $16,000 worth of conveyer equipment would be required. So, the total cost would be $571,000.

Although estimated costs for the two proposed robotic systems seemed high, they were within reason for Central Electric Supply. Other pieces of equipment used in company operations had at times cost some half-million dollars.

ROBOT VERSUS MAN

However, a cost/benefit analysis yielded some significant results. The manual deburring operation required workers paid at the rate of $6.35 per

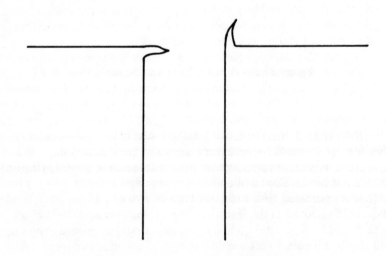

Figure 21.3. Two types of burrs.

hour, plus approximately $4 per hour in benefits. Furthermore, $960 per year was expended in the repair of hand tools and the purchase of expendable items such as gloves. The total cost per part for the deburring averaged 1.7¢ in labor and material. The scrap rate was one part in 200. The internal value ascribed to the part as it entered this operation was 68¢. So, the potential savings in labor costs was calculated as shown in Table 21.1.

It would be necessary to produce between 70 and 85 million parts to break even with the cost of the robotic equipment. To compute the pay-back period of each system, the equipment costs were divided by the average number of parts to be produced per year, 1,200,000. Pay-back would be 57.843 years for the nonvision system and 69.975 years for the robot vision system.

Clearly, it was not feasible for robots to do this particular operation. The price of labor was too low, the value of the part too high considering the existing scrap rate, and the task complexity and judgment level were too great for a practical robot system.

MAN IS BETTER

A lesson can be learned from this example that what human beings can do virtually without thinking about it, machines at the current state of the art cannot do without a great deal of effort. It is a simple task for a person to look over a wide area and select the particular topographical feature that must be altered. It is a complex task for a computer to visually scan the same area and make the same judgments.

The internal mapping characteristics of the human brain, allowing us to compare an image we have seen long ago with another that is at hand, allows us to make sophisticated and complex judgments. The current state of the art of a vision system allows for a much smaller number of visual components to be analyzed, typically 512 x 512. And a bit-by-bit analysis of these, even using some mathematical algorithms for decision making, requires at computer speed much time and great expense.

Central Electric Supply, rather than purchasing any robotic equipment, added a second human worker who could relieve some of the work load and reinspect parts for additional needed deburring.

Table 21.1. Robotics Systems Savings and Pay-back

Savings

$$\frac{68\cancel{c}\ \text{cost of part}}{200\ \text{reject rate for human}} + 1.7\ \text{production cost} -$$

$$\frac{68\cancel{c}\ \text{cost of part}}{50\ \text{reject rate, vision}} = 0.68\cancel{c}\ \text{savings/meter}$$

Pay-back Period

Vendor 1

$$\frac{472,000}{.0068} = \begin{array}{l} 69,411,764 \\ \text{parts to break even} \end{array}$$

$$\frac{69,411,764}{1,200,000} = 57.84\ \text{years}$$

Vendor 2

$$\frac{571,000}{.0068} = \begin{array}{l} 83,970,588 \\ \text{parts to break even} \end{array}$$

$$\frac{83,970,588}{1,200,000} = 69.975\ \text{years}$$

CHAPTER 22

THE COMPLETE MACHINE

A robot is often considered an isolated piece of equipment performing a certain task within the manufacturing realm. Increasingly, however, the robot is only a part of a larger automated system. The robot works in conjunction with other machinery to form a manufacturing cell or even an integral manufacturing unit for the production of goods. Currently, there is much debate regarding the proper use of robots and other automatic equipment and how integrated the manufacturing process can become in the future. Terms such as *flexible machining systems* and *factory of the future* are used to describe the interrelationships between a variety of machines performing diverse tasks.

A complete machine is now capable of being designed and produced by machine with little or no manual work required.

A COMPUTER SCREEN

Figure 22.1 shows the first element in the complete machine package: a computer screen capable of aiding the design process. Referred to by the initials CAD for *computer aided design*, the screen and related equipment allow a designer to implement and test ideas quickly without resorting to hand drawing or actual prototype construction.

A sophisticated computer program receives information from the designer and plots points on the computer screen. The designer sees two- and three-dimensional images of the part to be produced. Changes can be made on the screen to allow minor variations in the product's construction. CAD machines currently are used in great number. Usually, the next step is to have the CAD machine print a hard copy of the design drawings and to have this hard copy interpreted by humans for production.

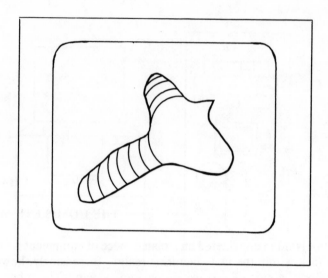

Figure 22.1. A computer aided design (CAD) unit allowing the
generation of a part design on a television screen.

CAM EQUIPMENT

However, with the second element in the complete machine package,
computer aided manufacturing (CAM) equipment (Figure 22.2), a direct
transfer of design from one machine to another can be accomplished along
with the automatic programming of machine tools for production. The
sophistication of computer aided manufacturing equipment is such that a
selection of raw material from bins can be automated, as can each step of
the manufacturing process through packaging of a finished component.

A ROBOT

To further the manufacturing process, a robot (Figure 22.3) is often
necessary to transfer parts from one set of machines to another or to
manipulate the parts in a way the machines cannot. The robot must interact
with the CAD and CAM systems and be reprogrammed by them for
different tasks, as needed. There are some robots now marketed that have
the ability to communicate interactively with other computers and to
change their programs spontaneously. Some standardization is required so

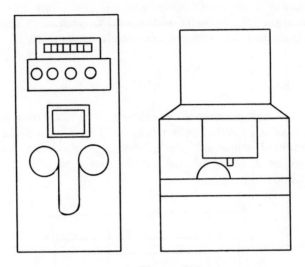

Figure 22.2.   Computer aided manufacturing equipment (CAM).

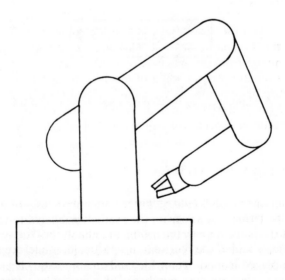

Figure 22.3.   A typical robot is also a part of the complete system.

that different makes of equipment can be used by a common communication link. Industry-wide standards such as the RS232-C allow many manufacturers to produce compatible equipment.

FEEDBACK

For the machine package to be complete, the system must accomplish a feedback task at the end of its loop (Figure 22.4). This might be the storage of documented materials about what was produced, exact design parameters or a listing of the number and type of each component manufactured.

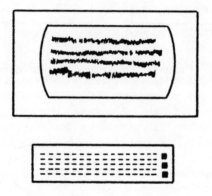

Figure 22.4.   A computer may receive data as a feedback of the entire process.

PROBLEM SOLVING WITH CAD/CAM

The advantage of such a complete machine package can be seen in the following scenario. Let's assume the product produced is an automobile. The warranty claims department of the manufacturer receives complaints about a squeak behind the rear seat in a particular model automobile. The reports are placed in a computer file, and an automatic checking program notices a more than random number of this particular complaint.

After assignment of the problem to a human engineer, it is determined that a change in the position of spot welds in the manufacturing cycle should eliminate the squeaking. Using the CAD equipment, the engineer analyzes the problem causing the squeak and constructs a suitable alternative pattern for the spot welds. The CAD equipment communicates a change in the manufacturing process to the CAM equipment. The CAM equipment passes along the message to the robot controller which repositions the welder to the new locations.

Meanwhile, a footnote is placed in the memory files of all the company dealerships indicating that automobiles produced after a certain date will have welds in slightly different positions. The information loop is complete when a notation is made in the memory bank used by the warranty department. This notation signals the complaint as potentially solved and requests an update should any warranty claims be made for the same problem on later model vehicles.

TRADITIONAL PROBLEM SOLVING

Compare this system of making a manufacturing change with the traditional one: As thousands of pages of paperwork arrive at company headquarters each day, it is unlikely that a complaint about a squeak would receive much attention. Among the thousands of complaints about noises, a regular pattern might not be noticed by human readers. Indeed, no one person would read all of the complaints to be able to see patterns developing. But let's assume some unusual circumstance, such as a high number of the same complaints originating at an individual dealership, causes the problem to come to light.

The problem is assigned to an engineer. To have a working knowledge of the problem, the engineer has to make physical inspection of a squeaking car. Assuming that one can be procured for his study, he spends much time investigating the nature of the squeak. Once the problem has been found, drawings are necessary, along with high-level mathematical analyses to develop a suitable change in weld structure. All of these activities—the inspection, the analysis and the designing—require much more time in comparison to the same operations done on computer.

Next, the engineer must institute an operational change at the manufacturing level. Drawings in hand (signed by the proper authority), the engineer makes his way to the manufacturing plant and shows the prints to the welding foreman. The foreman, comparing the new prints with the old ones, agrees to make the change and undertakes a retraining program for each of the human welders. With three shifts, three different foremen and three sets of welders including relief personnel have to be trained in the new weld positions. Depending upon the workers' desire to make the part better and their ability to accurately position the welder to the newly established points, the task may or may not be successful.

A letter and drawings outlining the weld changes should now be sent to the thousands of dealerships. However, this task, because of its expense, might never be undertaken.

Interestingly enough, the feedback loop described for the complete manufacturing unit might be accomplished as easily without the computer as with it. That is, once identified, a particular problem could be looked for by personnel examining warranty statements to see if it still existed after the change date. Although the manual process would take more time, the feedback result should be the same.

TIME COMPARISON

A complete manufacturing system could be expected to make the weld changes as outlined in just a few hours, once the need for change had been established. The manual system quite easily could require several months for the same task. And in fact, the job might never be accomplished at all. How many more squeaky cars would there be on the road?

QUESTIONS

1. What is CAD equipment, and what benefit does it offer?
2. What is CAM equipment, and what is its advantage?
3. True or false? A robot can interact with CAD and CAM equipment and be reprogrammed by them for different tasks.
4. In a complete machine package, what happens in the *feedback* step?
5. True or false? A complete machine package often can diagnose and solve manufacturing problems much faster than can human workers.

CHAPTER 23

BECOMING FAMILIAR WITH ROBOTS

The following projects have been designed to help students become more familiar with robots and develop the ability to use them to solve manufacturing problems. The information presented is purposefully sparse—details can be filled in by the student.

PROJECT 1

Design a robot installation using standard robotic equipment that will perform the assembly operation of a robot arm. Figure 23.1 shows the aluminum casting comprising the greatest portion of the robot arm. To this, bearings (Figure 23.2) must be added, spacers (Figure 23.3) placed between bearings, and the shaft (Figure 23.4) pushed through the assembled elements.

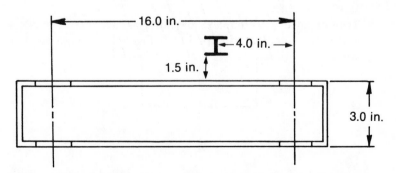

Figure 23.1. A casting in position for machining with an obstruction.

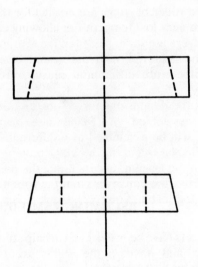

Figure 23.2.   Bearing to be inserted into casting.

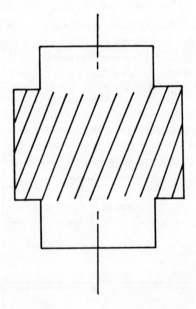

Figure 23.3.   A metal spacer.

Four of the tapered roller bearings are needed for the assembly of this arm, each cooled to a very low temperature allowing easy placement and pressed fit. Two spacers are to be placed in the arm, one at each end. A very close fit can be achieved with the seated bearings. The shaft, completed by a temperature differential press, must be handled and inserted very quickly.

Figure 23.1 shows the dimensions for an obstruction that will be very near the arm as it is assembled. An I-beam close to the arm cannot be moved, nor can the arm be assembled at a different location. Another design limitation is the placement of a moving conveyer on the side of the robot arm opposite to the I-beam, necessitating the positioning of a robot either completely across the conveyer from the robot arm or next to the beam. In designing the installation to assemble a robot arm, be sure to include:

A. A rough design of the gripping tools necessary to handle the bearings, spacers and shaft. (The cold-pressed components must be handled only with ceramic-coated materials to preserve their low temperatures as long as possible and to protect the gripping tool from damage by repeated super-cooling.)
B. A design of the floor layout necessary to supply each of the parts to the robot. (This does not include the design of the supply components themselves.)

In designing the installation, consider the following:

1. What types of robots could be used for this job (LERT classification—linear, extension, rotation, twist)?
2. What specifications could be given to robot manufacturers?
3. In human language, write a program for this operation.
4. If the robot is able to travel at three feet per second, and takes one second to accelerate and decelerate (no matter how fast it travels), how long will this operation take?

5. Is there an advantage to using two robots?
6. If you are in a classroom situation where others are working on this project, trade specifications with someone. Each of you write quotes from three imaginary vendors whose robots will meet specifications and accomplish the task in different ways.

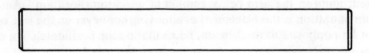

Figure 23.4.   The shaft that will complete the bearing assembly.

PROJECT 2

An additional project can be done based on something right in front of you. Imagine that you formed the U-Build-It Company and can manufacture any product available to you now. It might be a desk, a chair or a lamp, but it must be something now in your presence that you can observe closely. List the various manufacturing steps necessary to create this particular item from primary raw materials. Consider how many different operations in this manufacturing scheme could justify the use of a robot. Then develop a scheme for producing this project at a low price with robots.

If possible, you may wish to contact a manufacturer of the item under consideration and ask specific questions about manufacturing techniques and how they relate to robots.

PROJECT 3

For this project, consider a group of people most of whom are employed. This might be the robotic students in your classroom or your co-workers at your place of employment. Compile a list of the different jobs these people hold. Considering today's technology, decide which of these jobs might be done by some form of robotics. What percentage of the total number of jobs fall into this category? Remembering that from the invention of the microprocessor chip in 1971 to the placement of almost one million computers in the hands of the American population took only 10 years, which of these jobs conceivably could be done by robots 10 years hence?

PROJECT 4

Develop a program for a mobile and articulated home robot to perform a task such as the heating of canned soup. Assume that the robot is on the kitchen counter. A pan, a can of soup and a water tap are all available. The burner of a stove is accessible to the robot and may be started by it. Can the robot be made to follow a command to make soup? Hint: it's not easy.

CHAPTER 24

NONINDUSTRIAL ROBOTS

Robots have been used in manufacturing facilities for many years, doing the type of work that is both repetitive and easily programmed. In recent times, however, a new class of robots has emerged to serve other needs. Some now exist; some are in the experimental stage. These robots can be called nonindustrial and fall into four distinct categories. These are (1) the home or hobby robot, (2) the teaching robot, (3) the toy robot and (4) the commercial or service-type robot.

ROBOTS FOR THE HOME

A home robot's task is not production. Rather, it should be able to perform a variety of tasks in different areas of the house, and, while performing these tasks, it should provide some entertainment or educational value.

The requirements of a home robot are in some ways more sophisticated than those of most industrial robots. It would have to have some type of universal grasping attachment that could be used on a wide variety of items. In comparison, the industrial robot usually has a grasping mechanism suited only for a single product or family of products, each item similar to the others.

The home robot would have to be mobile—able to travel throughout the entire house doing its assigned tasks. Most industrial robots, on the other hand, are fixed to a single location within the manufacturing plant, being able to move perhaps only a few feet on a transfer mechanism.

The home robot would have to have a personality and have the appearance of a creature or intelligent entity. The most popular types now being produced have a voice provided by a voice synthesis unit, usually built into the robot cabinet itself. Industrial robots usually are constructed in a purely functional form; a lifelike appearance or personality is deliberately avoided. This is to help eliminate the perceived threat of worker attachment to or alienation from the robot based on personality.

The home robot would have to be weak, slow and small, so that it would be safe to be near, even during a malfunction. If a home robot is to place something such as a newspaper in the lap of a person, and the arm malfunctions, the arm would have to be weak enough to avoid hurting the person. The robot would have to be slow enough so that it would not smash anything valuable or cause harm to nearby humans. It would have to be small and light so that if it fell over on someone, it would not cause injury.

In order to maintain cleanliness and to operate practically in a standard household, the robot would have to work electrically and have no exposed lubricated parts.

The current hobbyist market has supplied a number of robots meeting these requirements (Figure 24.1). Because prices are now in a reasonable range, the market is growing.

There are many household tasks that these robots (Figure 24.2)—given the correct programming—might perform. Among these are: picking up the laundry, removing clothes from the dryer and putting items in drawers, vacuuming the carpet, opening a canned beverage and taking it to someone, and answering the phone.

None of these tasks is so burdensome that the healthy person need resort to a robot to have it performed, and none is sufficiently difficult to warrant the economic expenditure. Yet, the novelty and experience of the robot activity might make it worth the price. If we examine each of the above mentioned possibilities, we see that there are some similarities to industrial applications.

Picking Up Laundry

A small home robot equipped with sensory mechanisms could find laundry in several different ways. Once the robot had mapped out a room for the location of furniture and other objects, it could, at an appropriate time, scan the entire floor area for loose pieces of laundry strewn about. It might have a sensor on its base that would indicate when the robot struck loose clothing. Or it might have a visual detection system that could recognize the contrast between the clothing and the carpet. Another possible sensor would be one working off of infrared radiation that might, for example, pick up black socks due to a greater heat content, assuming they were placed in a sunlit location. When the robot located articles of clothing, it would lower its arm, close its gripper, raise its arm, check for

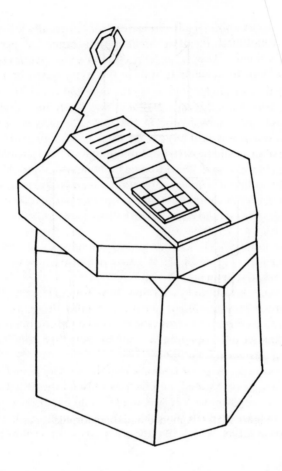

Figure 24.1. A small mobile robot available to the public.

attainment of the article and take the article to some programmed destination like a laundry hamper.

Clothing that was pushed under a bed, tossed into the back of a closet, thrown downstairs or otherwise inaccessible would not be picked up by the robot maid. For such random placement, a human-type servant would be necessary. If a person could be trained to throw soiled socks only on the portion of the floor reachable by a robot, he also could be taught not to throw them on the floor at all!

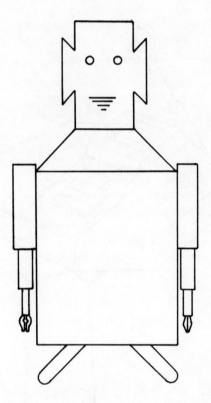

Figure 24.2. A mobile anthropomorphic robot.

Putting Away Laundry

The current state-of-the-art of home robots does not allow them to sense and sort clothing by type. However, an automatic dryer might be unloaded by a robot. That is, the door could be opened and all of the clothing removed, and either all of the clothing or those pieces easily identified by color could be put in a specified drawer.

A scheme to allow the robot to do all of the clothes sorting would require a human to load the washer and dryer on a regular schedule. For instance, slacks might be washed and dried one day, shirts another day and underwear a third day, allowing each item in turn to be placed in its proper drawer automatically.

This would place restrictions on the normal household activities and probably would increase the amount of absolute work. However, the novelty of the robot, plus the advantage of the dryer being unloaded before clothing wrinkles formed, might make it worthwhile.

Vacuuming The Carpet

The small home robot would have difficulty pushing the normal home vacuum cleaner. The drive motors in the robot's wheels would be strong enough to push the robot, but not a heavy machine. Yet a lightweight vacuum cleaner, particularly one similar to the type used in automobiles and plugged into the cigarette lighter, could be held in the arm of the robot and guided over the entire carpet surface area. The robot could also be programmed to vacuum other items reachable by the robot such as chairs and ashtrays.

The vacuuming robot could be programmed to perform its task when household members would not be disturbed. For example, if everyone were gone between 9 a.m. and 5 p.m., the robot could do its vacuuming then. Good programming, however, might have the job starting at 4 p.m. and ending just before someone came home. The reason for this is that should the robot malfunction and get stuck on something, the robot would not be vacuuming one spot in the carpet for eight continuous hours.

Opening Beer Cans

It might be said that if someone must drink beer, he should at least undergo the exercise of getting it himself. However, a robot servant that would upon command retrieve beer or some other beverage from a refrigerator would be of particular usefulness during football games or important golf discussions. It also would provide amusement to guests.

The process of opening a refrigerator door may seem like a simple one for a robot: merely grasp the door handle (if it is low enough) and pull backwards. However, the robot's ability to move with only a few ounces of forward thrust might not allow it to open a heavy door. Furthermore, the shelves of a refrigerator are oftentimes deep, and a small robot's limited reach might not be sufficient to retrieve anything other than what is on the front of the shelf. For this reason, some type of inclined rack allowing new cans to roll forward and a door with some self-opening feature might be necessary for the home robot to perform its task.

Answering the Phone

The annoyance of a ringing phone often leaves one wishing this task could be delegated to someone else. A robot servant with a voice synthesizing unit could answer a phone, provided it was within reach, and deliver a predetermined message to the caller. However, the robot would require considerable programming. What's more, if the robot were engaged in some primary task like vacuuming, the task would be delayed by the phone interruptions. It would be easier and less expensive to install a phone-answering machine.

Possible Future Tasks

There are many things that the current home robot cannot do that would be highly desired by potential owners. Among these are washing dishes, scratching a back, walking or washing the dog, protecting from burglars and making a 10-to-one martini (with or without olives).

The reason these tasks cannot be performed adequately is the robot's lack of sufficient sensory and manipulative ability. However, as industrial robot technology filters down into the realm of the home robot, each of the above might be a viable robotic task.

TEACHING ROBOTS

Teaching robots are a separate category only because of price. In general, all of the characteristics of the industrial robot are found in the smaller, less expensive teaching robot (Figure 24.3).

For engineers, programmers and repair people to learn robotics, they must have real hands-on experience with robot components and processes. And because of the high cost of industrial robots, particularly those using servo-control mechanisms (even a low-priced industrial servo-type robot will ordinarily cost over $30,000), it is not possible for many schools or individuals to use them. The teaching robot costing perhaps a few thousand dollars is a viable alternative in that it supplies the same type of components used on its more expensive counterparts.

A teaching robot usually is supplied with small electric motors to manipulate its arm system, an amplifier drive package to change computer-type signals into sufficient energy to drive the motors, a series of mechanical linkages constituting the mobile robot arm, and a computer interface allowing the study of the computer functions robots employ.

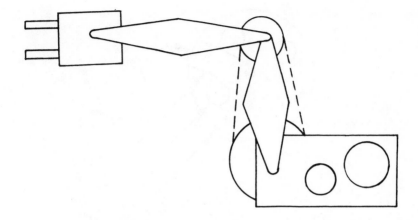

Figure 24.3.   A small educational-type robot.

Because the components are smaller and of a different quality level than those used in industrial robots, their price is much lower. Oftentimes, the computer is not actually a part of the robot and can be purchased separately. The robot also can be connected to a computer already owned by the purchaser. Sensory equipment for the educational robot can be made inexpensively since a high level of accuracy and dependability is not needed. Figure 24.4 shows a simple form of rotary encoder that can be built for a teaching-robot arm. The cost of only a few dollars is needed to produce this equipment, as opposed to many hundreds to produce its more ruggedly built industrial counterpart, with perhaps one hundred times the accuracy.

The home computer signals that drive the robot's motions are usually configured along the lines of a standard output signal. For example, a connection on the computer that ordinarily would go to some type of printing device would go instead to the teaching robot. A demand such as print would be interpreted by the robot as a command for some action. A few sets of standardized words and symbols would then be the programming format for the robot to move. For example: print "GET 5" might mean to move the fifth axis forward. The command print "GO 5; 71" might mean move the fifth axis 71 counts of the encoder and then stop.

The teaching-type robot need only be able to lift a few ounces and perform simple tasks to demonstrate the ability of robots to perform a wide variety of tasks. In a college setting, these teaching robots are often put to sophisticated and very demanding use.

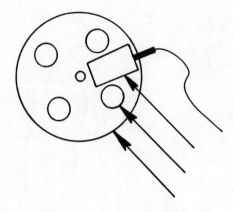

Figure 24.4.   An optical encoder.

TOY ROBOTS

The toy robot (Figure 24.5) is not much different from the radio-operated toy cars popular a few years ago. A simple control module causes the robot to move about and sometimes perform a simple task. These are somewhat amusing to the controller; but, since they do not replicate the values and features of more complex robots, they are not an educational tool. And, since they do not have a personality, they are not as interesting to the user or observers. An examination of the available toy robots indicates that for the most part they are almost as expensive as their more realistic counterparts.

COMMERCIAL OR SERVICE ROBOTS

Robots that are of a practical and working type and yet do not fall into the industrial category can be considered to be commercial or service-type robots. These in some ways might be more valuable than industrial robots.

Robot Guard

It is possible to build a robot guard. Many companies throughout the world make a product similar to a human guard, yet entirely robotic. These guards are now about at the sophistication level of a highly trained guard

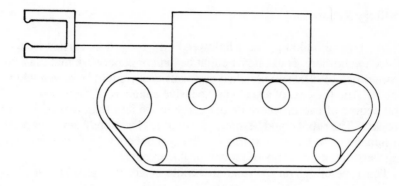

Figure 24.5.   A toy robot.

dog but are being given greater capabilities each day.

Consider the criteria for a good human guard. We might expect the guard to be alert, sensitive, mobile, able to disable intruders and summon help. The guard must be secure against attack and safe for those legitimately in his presence.

Those same criteria can be met by a robot guard. It is constantly alert in that it is not distracted by radios or TVs and never needs sleep. A robot guard's sensors are more accurate than those of a human being; it can be alerted by signals that a human being cannot sense. A robot guard is mobile in that it can roam about the protected facility at will. A robot guard is able to encounter intruders and prevent their escape. (Legal limits probably will be established in the future as to the strength and potency of guard robots.)

A robot is able to summon help and can in fact do so while investigating some intrusion. A robot is secure against attack beyond the capacity of any human being, even to the point of being invulnerable to bullets or physical blows.

However, with a guard robot, there is always the danger that, upon malfunction, the robot would consider any person found at the facility to be an intruder and respond accordingly. An ideal situation results when a combination of human and robotic guards are used.

For example, a large facility might have a centralized command post staffed by a human guard and a crew of robotic guards patrolling the facility, each guard sending back information via television cameras. Unlike stationary cameras, the cameras on the robots could not easily be bypassed or disabled.

Delivery Robots

In a typical building, many lightweight items must be carried from one place to another. These items might be mail or paperwork that must pass from office to office. Such delivery tasks easily could be assignable to a robot. And the cost of a small inexpensive commercial robot would, over the course of even a year, almost certainly be less than that of a human worker. The robot could be directed to various locations from some master command post. Containers in the delivery robot could lock, thereby allowing very secure transfer of money or documents.

This type of delivery system also could be adapted to such other functions as picking up trash along roadways. As mentioned before, however, as soon as a robot is released from a confined environment or allowed to perform a task of heavy weight lifting, there is a risk of human injury.

Other service-type robots will be discussed in the next two chapters.

QUESTIONS

1. List four categories of nonindustrial robots.
2. Give two advantages of a household robot.
3. List four ways a home robot would have to differ from an industrial robot.
4. What is the advantage of a teaching robot, and how does it differ from an industrial robot?
5. True or false? Robot guards are now about at the sophistication level of a highly trained guard dog.
6. Give three possible advantages of a robot guard over a human guard.
7. Give two advantages of an office delivery robot.

CHAPTER 25

SMALL WONDERS

There is a trend in robotics toward manufacturing components and assembling robots that are increasingly light in weight and small in physical dimensions. This trend is largely attributed to the availability of smaller materials and the realization by robot designers that large robot configurations are not necessary to perform many tasks.

Figure 25.1 shows a robot—not currently being manufactured but which could be in the future—for manufacturing watch components. If the entire operating range of the robot would be no larger than the watch, clearly the robot itself need not be large. It could be placed extremely close to the watch, indeed upon the same piece of fixturing equipment.

ADVANTAGES

An advantage of such a small robot would be in its controllability. Under current technology, a robot has a fixed smallest unit of motion based on its total motion. If a robot's arm movement can be divided into 1,000 points and the arm can reach 1,000 inches, the smallest increment of motion available to the robot arm is one inch. If, however, a robot's entire movement range would be one inch and this range could be divided into 1,000 units, 1/1,000 of an inch of accuracy would be achievable.

Another advantage of very small robot equipment would be the ability to place it in a small space, for instance, on a part of an assembly line where there was very little room for additional equipment. Inspection or assemblage of components could take place close to other operations already in progress.

185

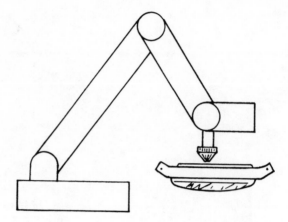

Figure 25.1.   A very small watch-making robot.

HEALTH BENEFITS

As a smaller example, Figure 25.2 shows an intravenous robot that might enter the human blood stream, approach obstacles such as blood clots and eradicate them. There are internal probes being used today that perform similar tasks. However, current equipment falls short of microfine robots that could perform sophisticated procedures. A robot might repair splits in arterial walls or remove obstructions such as minute tumors or accumulations of fatty deposits.

Such a robot would have to be connected to the outside world by a trailing power supply and hollow tube capable of removing from the body the intercepted and damaging parts. The manipulators on the end of the robot could be varied enough to cut, repair and cauterize any body part that they were capable of reaching.

The robot might also be able to perform such techniques as heart valve repair without surgical displacement of the heart or an excess of trauma, prenatal surgery with a minimum of disturbance to the unborn or mother, nonsurgical biopsy of tumors and the removal of physical manifestations such as kidney or gall stones.

An analysis of the human body reveals that much of it is hollow. What to the layman appears to be a solid mass of flesh and bone is filled, in the region of the torso, with air and food-process networks and, in the muscle regions, flowing fluid. Thus, much of the body volume would be available to a microrobot.

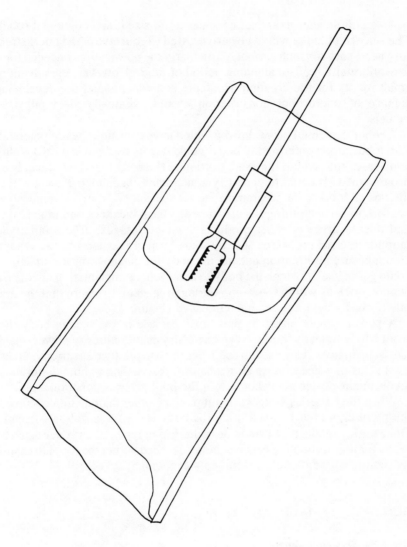

Figure 25.2.   A robot arm inserted intravenously.

As a still smaller example, consider a pill-sized, self-contained robot. The science of today would be hard pressed to conceive of the microstructure necessary for such a robot, particularly the power supply needed for it to work over a realistically long period of time. However, the advent of small robots for the creation of minute components and the developing science of microrobot construction should eventually yield pill-sized models.

A robot that could be swallowed would have enormous health potential. The robot, upon entering the body, could use its own navigational ability and senses to travel to the exact portion of the body for which it had been programmed. It could temporarily attach itself, becoming a guest within the body. It could be programmed at regular intervals, or continuously if needed, to do such things as measure chemical balances within the body and release drugs or other products in exact dosages. The robot could monitor internal processes and, by some form of telemetry such as radio, communicate information outside the body for diagnosis or treatment. It could guide itself through the body to areas where implantation might be a benefit, such as within the womb during pregnancy to insure that the umbilical cord would not become entangled (Figure 25.3).

A power supply might be developed for use in the human body that would draw upon the body's own chemistry as the source of power, much like a chemical battery uses an acidic reaction to produce electricity. Intestinal acids or some biological mechanism responding to human hormone levels might power the robot within the body over its functional life.

When the batteries were worn out or some other malfunction occurred, the robot could release its grip on the body and pass harmlessly from it. Once retrieved, the robot could be recharged or repaired and once again be ready for use. A homing beacon could be placed within the robot to ensure its return, where it could continue its economic payback.

BUILDING A SMALL ROBOT

Different Design Values

In order to build a very small robot structure, a different set of design values would be necessary for use in formulas calculating force and frictional units. Extremely small moving components respond differently to each other than do those of "human size." Much of man's design technology and mathematics used in physics have been developed for the human-sized world. That is, things are built on a scale natural and familiar to human beings. When we say that an object drops at a certain known

Figure 25.3.   An intrauterine robot for prenatal monitoring.

speed due to the effect of gravity, we are assuming either a complete
vacuum in the laboratory or some type of air pressure consistent with
human experience.

A set of design values based on components almost microscopic might
take into account the different physical nature of these components. Such
a simple thing as calculating the amount of friction on a bearing journal, for
example, is completely different when the bearing is 1/64 inch in diameter.
The way lubrication affects the rolling characteristics, the viscosity and
lubricity necessary for the lubrication, and the smoothness and relative fit
of the components are all very different.

The human body is an example of how very minute components can be
assembled and work in a functional whole, yet obey different laws of nature
consistent with their own sizes. A single body cell nourished by the blood
stream has a cell wall permeable to oxygen. Were this not the case, the cell

would quickly die from lack of the proper nutrient. However, the whole body has to our scale a skin impermeable to air, allowing only the pores to exchange gases with the surrounding environment. Any person who has had stomach gas realizes that the body on large scale is not permeable.

New Materials

If a robot is to flex some manipulator, it may not be necessary to exert pressure of 100 pounds per square inch (PSI) upon one of the components. On a small scale, pressure differentials of a fraction of one PSI could be used to accomplish a great deal. Indeed, the flow of blood throughout the body is accomplished with a comparatively low differential pressure.

In order to use these small pressures, the friction between the two materials comprising the joint would have to be very low. The joint, however, might need a seal to prevent both leakage of components or chemicals to the outside and entry into the robot of biological contamination that might destroy the robot itself.

By use of the proper materials, these two seemingly opposite needs could be resolved. Perhaps the use of graphite or biologically produced fibers, or even a hybrid combination of materials produced by bacteriologically splitting genes, might prove the answer to the microrobot's needs in the future.

Different Power

Another engineering difference in building very small robots is that very small power consumption and supply components would have to be designed. Although the energy requirements of an extremely small robot would be low, the available space for energy storage would be very small.

Packaging energy in very minute quantities and within a confined space is a technology begun only recently. With the development of the microprocessor, it became both feasible and desirable to make electronic components mobile and self-contained. Operating at approximately five volts, components had small energy requirements, and sometimes available power supplies were too much for the functioning portion of a device. A watch battery is an attempt to make a long-lasting power supply consistent with the energy requirements and physical size of the device it will accompany. Yet, for a microrobot such as those we have contemplated, even this power supply would be much too large.

It is possible that as yet undeveloped techniques will emerge to supply power for extremely small uses. Such technologies as the encapsulation and slow degradation of fuel components might be used for robotic supply, as might phenomena not yet explored.

Assembly By Robot

Still another engineering difference in constructing very small robots is that the assembly would almost certainly have to be done by a robot. With the ability to manipulate small components by artificial means, that is, not by the direct manipulation of hands, man's ability to construct took a giant step downward in scale. We've heard of the artist who paints the Mona Lisa with astonishing accuracy and resolution on the head of a pin. However, this might be surpassed by a microrobot who, upon the commands of a human artist, would contruct a three-dimensional replica of Michelangelo's David, whittling away from the stem of a pin.

NEED MUST BE ESTABLISHED

A rationale for construction would have to be established before the research for such incredibly shrinking robots would be undertaken. Just as with large equipment, even the initial research and development for very small components would be costly. On the other hand, the modern industrial robot was first developed to do a single series of tasks and gradually became more sophisticated. As robots become increasingly smaller, so the process of shrinking robots will be furthered.

The feasibility of microrobots for health care would most likely depend upon a mass-production scale. Just as a medicine is not lucrative for a drug company until it is used by large numbers of people, microfine robots would not be economically justifiable until they could be produced in large quantities.

A stumbling block in the development of such robots for health care might be public reaction. Most people would be extremely leery of a mobile apparatus traveling—under its own control—through their body. Some people might relate this to the swallowing of a bug and not being able to get it out of their system.

What's more, there would be a certain risk associated with machine failure. Should the robot's assignment be to monitor insulin levels in the vicinity of the pancreas and then gradually release insulin-producing hormones, a life-threatening situation could develop if a malfunction caused all of the hormones to be released at one time or even in incorrect measure.

These drawbacks could be overcome, however, if there were tremendous health benefits to be gained by the use of microrobots. After all, a person facing alternatives of either experimentation with some radical technique or death would most likely swallow live salamanders if it might save his life.

SMALL NIGHTMARES

As soon as any new technique is developed in science, it is put to use in ways other than those expected by its inventors. The rocket, originally intended as an item of amusement, was used even in medieval times for warfare. As we contemplate the possibility of microrobots, we might speculate on their use for destructive purposes.

Certainly, crime and espionage are potential areas of use for small robots—even of a size produceable today. Most modern security measures for both our personal and economic protection are based on the assumption that our greatest enemy is another human being. Many sophisticated security devices cannot detect things that are smaller than human size, or they are designed to sense and then overlook them. A blanketing system of sonic beam, for instance, that could sense people moving about in a warehouse might be tuned so that insects or rodents would go undetected, thereby preventing false alarms.

A robot the size of a squirrel or mouse easily might pass into a protected area and perform whatever function its programmer deemed appropriate. We could well see new spies in the form of squirrels secreting themselves in secure locations and, if detected, destroying themselves. A squirrel might enter a restricted office, climb up to a filing cabinet, open a drawer and take pictures of sensitive material, even with people around in nearby offices. An open window or even a ventilation shaft could give a squirrel-sized spy robot easy access to an office.

A visitor allowed within a building up to a certain point but denied access to restricted areas could carry the squirrel in a brief case. At the appropriate time, the squirrel could be released to find its own way—under furniture and without being observed—to a designated area. Terrorists could use such a device to carry explosives to virtually any place desired.

It is common to refer to a wire tap as a *bug*. In the case of a microfine robot, it might very well resemble a bug—perhaps a praying mantis—that could secrete itself within a communications area and tap directly into phone lines. Even a spy-squirrel could position itself upon a telephone pole outside a dwelling or office, appearing to be a legitimate member of the rodent family, and tap into phone lines.

To fan the flames of our paranoia, let us consider a robot snake able to slither through sewer pipes. Because it would not need to breathe, it could spend any time necessary to locate the drainpipe from a particular bathroom. It could swim up the pipe, come out of the sink drain, find a designated safe, enter a small vent hole in the safe, and seek out the material it was programmed to find.

Our same snake could be carried in a briefcase by a bank customer and placed within a safety deposit box. After the bank had been closed and the vault locked, the snake could drill from inside the box, removing its own lock, and push the drawer open. It could search other safety deposit boxes or any particular one chosen by the programmer by using its forked tongue with specially designed prehensile appendages that could approximate the shape of keys. The snake could swallow up jewelry, cash, stock certificates or other documents, lock the boxes, and return to the customer's box. Then, it could insert a replacement lock with the same geometry, swallow up the ruined lock mechanism and close the drawer.

The next morning, our customer could enter the bank, request access to his safety deposit box, and leave with snake and newfound wealth, creating a baffling mystery for some great detective to solve.

Fortunately for the safety of our society, we need not be too worried about microrobots performing the contemplated deeds. They are probably a number of years away and most likely will be very costly. But then again, what was considered an impossibility just a few years ago —a pocket-sized computer—is in fact now a reality.

PROJECT

Use your imagination. See how many potential functions you can come up with for small robots—pill size to squirrel size.

CHAPTER 26

GIANT ECONOMY SIZE

Just as robots can be made very small to accomplish feats beyond human ability, they can be made very large to likewise accomplish tasks impossible for human beings (without the aid of machinery). Robots are being designed that can lift heavy weights or large quantities of materials, reach deep into the ground or high into the air, or even travel into space, such as the robot used by NASA to unload the cargo bay of the space shuttle.

With large robots, the emphasis is on strength or reach, rather than on agility or speed. We begin to notice when robots become very large that they are less an artificial human being or a worker replacement and more a normal piece of equipment with robotic components. Generally, any machine can be robotized and given functions that are "automatic and programmable." A given piece of equipment might need a redesigned operating control system to accept a robotized function, but the base machine usually can remain very much as before.

ROBOT TRACTOR

A simple farm tractor could be made into a farm robot by the installation of (1) a computer-type control, (2) a servo mechanism capable of steering the tractor and working the gears, and (3) some type of sensory apparatus allowing the tractor to perform its work wherever needed. Such a scheme is certainly within the bounds of current technology. A control system no more complicated than one used in a standard industrial robot could be used to control a farm tractor in its direction, speed and work performance (plow, spade, till, reap).

On a typical tractor, there is sufficient extra electrical power to supply a computer control. And in many modern tractors, the cab unit is sufficiently

195

air conditioned and protected from the elements to allow a computer to function adequately. A servo mechanism very much like that used in a conventional robot could be used in combination with a type of power steering to turn the tractor's steering wheel, even against great force. The servo mechanism also could be used to automatically shift gears for speed changes or for the operation of the power take-off equipment.

Current robotic sensory units could be adaptable for tractor use. Either reflected beam sensors or sonic detectors could be located, for instance, between rows of corn to help tractors navigate. A Lorain-type triangulation system, operated by laser beacons placed at various positions throughout the field, could make possible an exact positioning of the tractor at all times.

Once the farm tractor was outfitted with its three robotic features, it could perform very much as it had with a human operator, with the exceptions of greater efficiency and reduced human labor.

Why does the farmer plow only during the daytime? Because the farmer, often the only person who navigates the tractor, becomes tired. So the production derived from the expensive tractor is reduced to the amount of labor the farmer is willing to perform during his daytime activities. Once robotized, the tractor could work 24 hours per day until the task assigned to it was completed, thereby increasing the work yield of the tractor and perhaps increasing the farm acreage that could be managed by the single farmer.

Figure 26.1 shows how a tractor might be redesigned to make more adequate use of robot features. Rather than having large wheels, a tractor

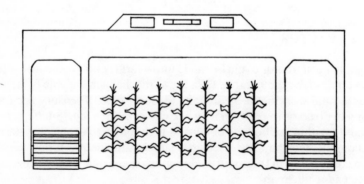

Figure 26.1.   A robot tractor.

could be designed with widely spaced caterpillar treads, thereby allowing a wide amount of uninterrupted field between the treads. Sensors could be placed in the field (where they would not be harmed by plowing), allowing the robot tractor to be self-guiding. Operations such as fertilizing, watering or weed removing could be undertaken while crops were growing.

Such crops as wheat, which are planted randomly rather than in regular rows, could be serviced the first time by the tractor without damaging the crop. Early in the season, weeds growing about the height of the wheat could be selectively removed, or at least trimmed, by the robot passing harmlessly over the top of the wheat itself.

At harvest time it would be possible to bring crops in in a relatively short time. Crops could be allowed to reach maximum ripeness, even in the face of an oncoming storm, because the harvest could take place on a continuous basis and at a much higher rate. A different scale of production could develop, allowing for a reduced cost and higher yield.

Farm machinery in recent years has become very expensive on the scale on which the average single-family farm is operated. Yet such a farm must be highly mechanized for its economic survival. The cost of a robot tractor system would not be higher necessarily than that of a modern-day multipurpose large-scale tractor. Indeed, if some manufacturer were to develop a robot tractor system employing the same economies of scale in production as current tractors, it could be lower in cost, even allowing for its robotic content.

MOBILE MILKING STATION

Another large-robot possibility—for use on a dairy farm—might be a mobile milking station. Almost all milk produced in the United States for public consumption is taken from cows by automatic milking machines. These are not robots as they must be connected by human hands to the cows. It is easy to imagine, however, a mobile milking station that would approach a cow, retrieve its milk and move on to do more of the same. (The cows probably would need specific training to allow a large milking machine to approach them, but this no doubt could be accomplished with proper techniques.)

The advantage of having the milking machine go to the cows rather than vice versa is not necessarily economy but convenience for the cows. In warm weather, it would not be necessary to keep the cows close to the barn, nor to herd them continually from prime grazing land to the barnyard, providing a truck was able to retrieve the milk periodically from the milking machine.

Cows could be broken into small herds and kept at some distance from the farm and each other. If a milking machine could handle 10 cows, those 10 need only stay in proximity to one another for the machine to be able to carry out its task. Certain diseases befalling one small herd of cows would not necessarily spread to the other herds. The automatic milking machine might even be programmed with many of the skills of a cowboy. It might herd the cattle, watch over them during the night and bad weather, and thwart rustlers and wild animals.

TANKS

An early theme in science fiction was the development of a completely self-directing robot soldier. While that might not yet be feasible, a robotic tank is (Figure 26.2). A conventional armored tank, much like the tractor, could be transformed into a robotized version. A modern tank already has many of the enhanced optical sensory abilities that a robotized tank would need. It has a sophisticated system to control speed and direction of treads and has on board complex equipment including microprocessing components of the type used in robots.

However, one of the largest deterrents to the production of a robotized tank is the potential for malfunction and loss of control. If a robot in a factory malfunctions and acts in a manner detrimental to itself or the surrounding equipment, power to the robot can be interrupted. But when a robot is completely self-contained (its power supply, control functions and programming all contained within itself), there is no way for a human operator to interrupt the power supply.

In the case of an unmanned tank made purposefully to resist attack, once it malfunctioned its potential destructiveness would be very great indeed. It might seem a simple matter to instill within the control system some extremely reliable signal that would cause all functions to cease within the tank. However, for the tank to serve its purpose, it would have to be protected from that very type of signal which might originate from an enemy.

MINING EQUIPMENT

Equipment used in modern day mines, either to retrieve ore or to set charges for explosives, is made on a scale compatible with human needs. Operated by a human being, the equipment must produce chambers in the mine that allow for human passage. Furthermore, there must always be

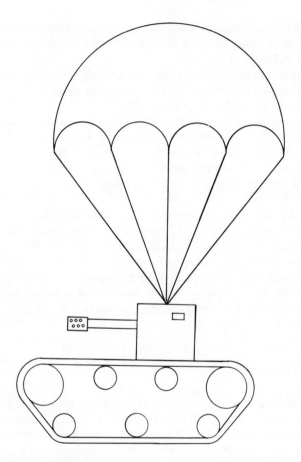

Figure 26.2.   Completely automatic robot tanks would not need crews.

supplied an oxygen level compatible with good health for the workers.

Robotic mining equipment would not be under the same constraints. The robot miner could be as large or as small as was necessary for the job at hand. For a coal mine that had veins of coal anticipated to be only a few feet deep, small-scale mining equipment could be used. For a mine of iron ore or other such material where voluminous veins are encountered, the mining equipment could be very large and remove great amounts of ore at a single pass.

The mining equipment would need a self-contained power supply. It would have to have a program feature allowing the seeking of ore rather than the following of a prescribed path. It also would have to have an ability to return to the surface, either at the end of its task, or at a command from the robot's operator.

Great interest has developed of late in mining the seas, that is, removing ore from ocean depths, where it is found in abundance and where rare materials can be located with relative ease.

Robot sea-mining equipment could be made on a scale approaching ship size. It could leave the ocean surface and travel with self-contained power for long periods of time, returning to the surface with large volumes of procured material. It could rendezvous with the ship where more energy could be installed and the recovered ore off-loaded.

In each of the cases of mining, there would be few humans subjected to danger during a malfunction; however, care would have to be taken that the very expensive equipment could be preserved from destruction.

One advantage to having very large equipment would be an increased ability to find it in the event of a malfunction. An enormous steel structure under the ground could perhaps be located through sound waves or metal detection equipment. A very large undersea mining apparatus could be found with sonar or, if it automatically returned to the surface, by an air search of the sea.

ROBOT BUILDINGS

Robots could approach a scale wherein they could be occupied by human beings. Even now, modern office buildings are being equipped with very sophisticated operating systems for heating, cooling, communications, internal transportation (elevators and escalators) and other services. Many of these systems can be considered as robotic. But when all of the systems are combined, the entire building takes on the attributes of being a robot in and of itself (Figure 26.3). In the not too distant future, all functions of a building may be robotized.

Currently, we have elevator doors that open at the push of a button and that will not close when someone is within the doorway. Why not have all of the doors of the building similarly fitted? If a personal ID card were worn, the door sensors could respond automatically to certain cards, allowing passage of authorized people into certain areas and restricting passage to others.

File cabinets also could be fitted so that a standard card would allow only authorized users to open the drawers. Care would be taken, of course, that in the event of some malfunction of the file cabinet security system, doors could be opened manually.

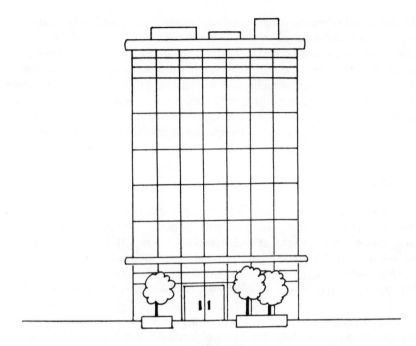

Figure 26.3.   An entire building may take on the characteristics of a robot.

Other building functions could be automated for ease and efficiency. It is assumed in modern society that every building of even a modest size has rest room facilities. Usually, however, other services such as food and package delivery are offered only in certain areas of a building, usually by human service personnel.

A completely automated building could have a centralized robotic kitchen facility with channels allowing immediate delivery of food, as commanded by the hungry person, anywhere in the building. The same channels could deliver small parcels or files to any place equipped to receive them. Such a conveyer system would be much more efficient than package delivery via human beings or mobile robots.

Building security during times of low occupancy also could be a function of the central robot system. The system could monitor hallways and access points and could sense potential catastrophes such as major water leaks or fires. A fire prevention system incorporating smoke detector alarms, an automatic sprinkler system and a communication link with the fire department could be automated. In the event of a fire, sprinkler heads, not only in the fire area but on the fringe, could be turned on automatically to alert occupants out of the range of smoke alarms.

The fire department could be notified before its arrival of the nature and location of the fire. Indeed, in a system incorporating a large number of television cameras for surveillance, robots could train cameras on portions of the building nearest the fire, projecting these images directly to the fire department. This would alert fire fighters to bring any necessary special equipment such as gas masks or special-purpose fire extinguishers.

PROJECT

In the classroom, have a brainstorming session on large robots. Here's how it works. Students come up with as many use ideas as possible for large robots—tractor size, ship size, or larger. One student writes them on the board. At this point, there is no attempt to analyze their feasibility.

After several ideas have been listed, the class breaks into three- or four-person workshop groups. Each group selects one idea and develops it.

Groups should consider such questions as: What is the robot's primary objective? Are there any secondary objectives? What are its advantages? Who will use it? What should it look like? Of what material should it be constructed? How will it be powered? What measures will be taken for human safety?

After a specified time, the class reconvenes, and each group reports on its robot.

CHAPTER 27

TRANSPORTATION ROBOTS

A popular fantasy image is of a vehicle driven by a robot. Usually, the robot is anthropomorphic and operates the vehicle as would a human being (Figure 27.1). Upon reflection, however, it appears likely that rather than a robot being designed to operate a conventional vehicle, a specialized vehicle will be developed with robotic functions an integral part of the machine.

An example of a robotized transportation vehicle is a shuttle bus connecting several locations on a predetermined route. Usually, no other traffic is allowed in the pathway of the bus. Much like an elevator, the bus allows people to enter, closes its doors automatically and safely, and begins its journey.

Other transportation devices lend themselves well to robotic operation. A high-speed train, allowing very little reaction time for the operator, would be well suited for robotically controlled speed and emergency features.

Figure 27.1. A car driven by a robot.

More sophisticated computer techniques and a greater understanding of reliability factors in robot design and construction allow robots to be used for increasingly complex transportation tasks.

ROBOT PLANES

Research has been done on creating a completely automatic jet airplane, with all navigational and pilot functions automated by computer. A modern autopilot system already uses many of the sensory mechanisms necessary to completely automate the plane. Indeed, the pilot of a large jet aircraft would be at a grave disadvantage if no automatic functions were at his disposal. Although the plane could be landed with only manual control, the situation would be, at best, one of great concern for everyone. To completely automate the plane, including takeoffs and landings, the autopilot functions only need be expanded to include short-range position detectors (for the exact position of runways, etc.) and preprogrammed instructions for aircraft destinations.

It is expected that once automated planes are fully developed, greater safety will be experienced by the flying public. However, it is difficult for passengers to give up the psychologically satisfying idea that there is always a human being ready to take over. (We all remember the story of the robot pilot announcing to its passengers, "Have no fear. Nothing could possibly go wrong—go wrong—go wrong—.")

We now have problems trusting equipment that succeeding generations will not. For an analogy, consider the early days of the automobile. Many thought the automobile was a bizarre and dangerous creation. There was speculation that women's faces would be permanently distorted riding in a car with wind velocities approximating 10 miles per hour. Some people thought that the automobile would never be as safe as a horse for carrying a passenger. After all, a horse would not voluntarily smash its rider into a brick wall as would an out-of-control automobile.

Actually, there was a much greater chance that a horse would lose its composure and throw its rider or gallop away pulling an out-of-control wagon than would the driver of an automobile lose control. What's more, ending the heavy reliance on horses as transportation resulted in secondary benefits such as a decline in diseases transmitted both by horses and flies.

ROBOTIZED AUTOS

Now, of course we trust the automobile, and it is our primary means of transportation.   Nevertheless, it's likely to be the last transportation method robotized. Even when statistics prove that an automatic system is safer than manual operation, there is likely to be great reluctance to give up control.   However, robotization no doubt will come. The sequence of events might be somewhat as follows:

At first, only some robotic monitoring functions would be added to conventional automobiles. The beginning of these are seen already—monitors at high-frequency intersections or freeway lanes signaling the operator: change lanes, accident ahead, detour and so forth.

Additional features would come with time, such as a close-range, object-detection system. This would signal the operator when a lane change should not be attempted because other cars were too close and when the automobile was following too closely the car in front. Current weather conditions would be considered. A series of lights or an audible signal would alert the operator to impending danger.

In our robotized vehicle, automatic speed selection would be regulated not to exceed the legal limits. As a special option, automobile manufacturers might provide a mapping system, automatically giving directions to the operator for any preselected points. Instructions fed automatically into the machine could find new locations, even some establishment that had recently moved. It might be possible to arrive at a destination by homing in to its telephone number. Or, each organization might have a set of location instructions—either printed in a telephone book or available over phone lines—to automatically direct an automobile.

The ability of a car to monitor highway conditions, respond to changing traffic and environmental considerations, and find destinations would lead the way toward development of a fully robotized automobile. For security, a set of manual controls might remain part of the car's basic package. But with time, much like parachutes on a passenger aircraft, a steering wheel probably would be considered an unnecessary encumbrance.   Why sacrifice leg room and freedom of movement for a steering wheel and gas pedal that are never used?

OTHER ROBOTIZED TRANSPORTATION

A robot vehicle with the ability to arrive at any selected destination without operator intervention would be the perfect chauffeur for children or package delivery service. Robot taxis could be relatively inexpensive to operate and could function solely with credit cards, eliminating robbery of drivers.   A robot cab would not hesitate to wait hours, should it be programmed to do so and paid for by its customer.

It could be programmed automatically to seek out high traffic areas at different times of the day. For instance, it could seek passengers at the airport in the morning, within the city during the day and again at the airport in the evening.

In addition to automobiles, trucks could be robotized. A truck could be loaded and dispatched to a location in the robot's memory. The robot would not need to stop for human bodily functions or food. The vehicle would be programmed to obey laws and adjust to traffic conditions. Any problems or standstills immediately would be detected by local police, to discourage highway robbery.

PROJECTS

1. Discuss the following.   You show up at the airport for a cross-country trip. You are attending a crucial business meeting, and this is the only flight. The gate attendant gives you a boarding pass and informs you casually that you are riding on one of the new robot-piloted planes with no human crew. What would you do?
2. Discuss how you would sell a robot-piloted airplane to the public if you were the marketing director of an airline introducing such a model.
3. Discuss how soon—if ever—we will be riding in robotized automobiles that we program.
4. Discuss the pros and cons of a robotized car to chauffeur children to school, club meetings, sports activities and other events.

CHAPTER 28

IN CONCLUSION

One of the main results of robotics and high technology is increased efficiency and a greater amount of goods produced at lower cost. A basic axiom of economics is: people spend what they earn. If it is possible using high technology to produce an automobile lower in price, the savings to the consumer will get spent on something else. And the amount of human labor saved in the automobile manufacturing process can be transmitted to some other product. Thus, there would be a greater absolute amount of goods available for a fixed unit of price.

The goal of our society should be to provide all people with meaningful jobs so that everyone can both contribute to and be serviced by the economy as a whole. Only with this goal in mind can the maximum social and economic benefits of high technology be achieved.

PROJECT

As a class project, review the following list of areas in which future expansion is predicted. Then discuss the possibilities for robot use in each.

1. Education—the distribution of knowledge.
2. Designing and equipping of buildings.
3. Entertainment—production and distribution of everything from live drama to video games.
4. Sports and related services.
5. Recreational activities—especially those possible in small units and available in urban areas to highly paid workers.
6. Social activities, particularly those of large groups such as company employees.

7. Science research, especially where the government provides a financial engine.
8. The arts—music, painting, sculpture, etc.
9. Administration—of machines as well as people.
10. Hand crafts.
11. Personnel services—particularly relating to retraining and replacement of the unemployed.
12. Health care.

Companies manufacturing robots in the United States are located primarily in three areas (Figure A.1). It is not surprising that these locations closely correspond to areas of high robot use and areas where similar machinery is manufactured.

Figure A.1. Location of robot companies.

COMPANIES AND THEIR ROBOTS

There are a multitude of robot companies in the modern marketplace. Some have been in business for several decades, while others—including some of significant size —are only a few years' old. The following are some of the most significant companies, plus basic information about their robots.

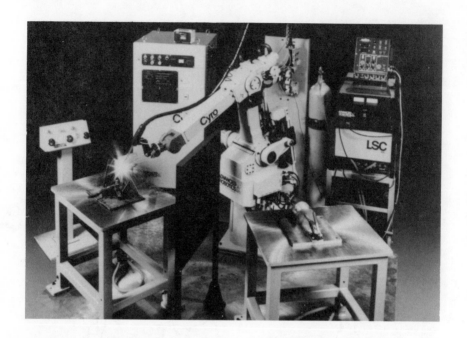

COMPANY NAME:	ADVANCED ROBOTICS
MAJOR BUSINESS:	Robots
COUNTRY OF ORIGIN:	USA
LERT:[1]	$L^3 R^2, R^4 T$
BASIC POWER SUPPLY:	Electric
PRIMARY USES:	Arc weld, machine loading
TYPICAL COST:	$80,000 - $100,000
TYPICAL ACCURACY:	$\pm.008$ to $\pm.012$ inch
TYPICAL WEIGHT CAPACITY:	Approximately 20 pounds

Photograph provided by Advanced Robotics

[1] Linear, extension, rotation, twist

COMPANY NAME: ASEA
MAJOR BUSINESS: Electric power equipment
COUNTRY OF ORIGIN: Sweden
LERT: R⁴ T
BASIC POWER SUPPLY: Electric
PRIMARY USES: Arc weld, spot weld,
                           machine loading
TYPICAL COST: $60,000 - $110,000
TYPICAL ACCURACY: ± .008 inch
TYPICAL WEIGHT CAPACITY: 12 to 200 pounds

Photograph provided by ASEA Inc.

COMPANY NAME: AUTOMATIX
MAJOR BUSINESS: Robots/industial machine vision
COUNTRY OF ORIGIN: USA
LERT: R⁴ T
BASIC POWER SUPPLY: Electric
PRIMARY USES: Arc weld, machine loading,
                         inspection, assembly
TYPICAL COST: $80,000 - $150,000
                         (machine vision)
TYPICAL ACCURACY: ±.002 to ±.008 inch
TYPICAL WEIGHT CAPACITY: 16 to 22 pounds

Photograph provided by Automatix

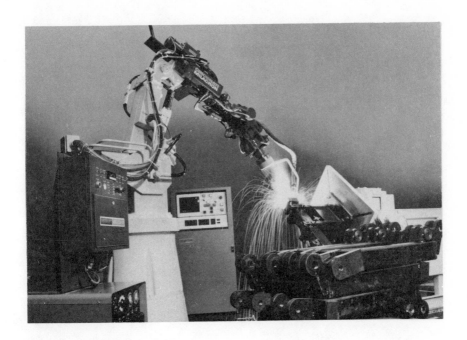

COMPANY NAME: CINCINATTI MILICRON
MAJOR BUSINESS: Machine tools
COUNTRY OF ORIGIN: USA
LERT: R^5 T
BASIC POWER SUPPLY: Hydraulic/electric
PRIMARY USES: Spot weld, arc weld, material
                          handling, forging, die casting
TYPICAL COST: $64,000 - $90,000
TYPICAL ACCURACY: ±.050 inch, hydraulic/±.006
                          inch, electric
TYPICAL WEIGHT CAPACITY: 14 to 255 pounds

Photograph provided by Cincinnati Milicron

COMPANY NAME: DeVILBISS
MAJOR BUSINESS: Robots
COUNTRY OF ORIGIN: USA
LERT: R⁵ T
BASIC POWER SUPPLY: Hydraulic
PRIMARY USES: Spray paint, die casting
TYPICAL COST: $100,000
TYPICAL ACCURACY: ±.08 inch
TYPICAL WEIGHT CAPACITY: 10 pounds

Photograph provided by DeVilbiss

COMPANY NAME:	GCA
MAJOR BUSINESS:	Robots
COUNTRY OF ORIGIN:	USA/Japan
LERT:	$L^3 R^3, R^4 T$
BASIC POWER SUPPLY:	Electric
PRIMARY USES:	Arc welding, spot welding, machine loading, material handling, assembly, inspection
TYPICAL COST:	$30,000 - $60,000
TYPICAL ACCURACY:	±.004 to ±.020 inch
TYPICAL WEIGHT CAPACITY:	10 to 2,500 pounds

Photograph provided by GCA Corporation

COMPANY NAME: I.S.I.
MAJOR BUSINESS: Air equipment
COUNTRY OF ORIGIN: USA
LERT: Components assembled
                           to fit each job
BASIC POWER SUPPLY: Air
PRIMARY USES: Material handling,
                           machine loading

Photograph provided by I.S.I. Manufacturing Inc.

COMPANY NAME: KUKA
MAJOR BUSINESS: Robots
COUNTRY OF ORIGIN: Germany
LERT: R⁵ T, L³ R³
BASIC POWER SUPPLY: Electric
PRIMARY USES: Spot welding,
                                 material handling
TYPICAL COST: $100,000 - $130,000
TYPICAL ACCURACY: ±.04 inch
TYPICAL WEIGHT CAPACITY: 200 to 1,000 pounds

Photograph provided by Expert-KUKA, Inc.

COMPANY NAME: MACHINE INTELLIGENCE
MAJOR BUSINESS: Robot systems
COUNTRY OF ORIGIN: USA, Japan (uses Yaskawa)
LERT: R⁴ T
BASIC POWER SUPPLY: Electric
PRIMARY USES: Inspection, assembly
TYPICAL COST: $150,000
TYPICAL ACCURACY: ±.004 inch
TYPICAL WEIGHT CAPACITY: 6 pounds

Photograph provided by Machine Intelligence Corporation

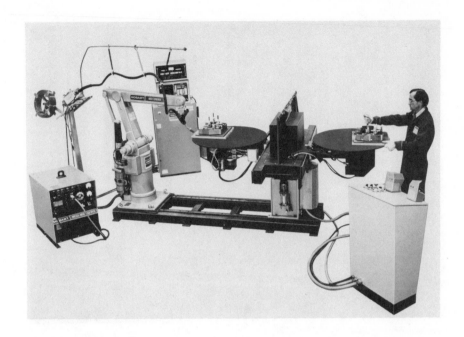

COMPANY NAME: MOTOMAN (YASKAWA)
MAJOR BUSINESS: Electric products
COUNTRY OF ORIGIN: Japan
LERT: $R^4 T$
BASIC POWER SUPPLY: Electric
PRIMARY USES: Arc weld, material handling
TYPICAL COST: $70,000
TYPICAL ACCURACY: ±.004 to ±.008 inch
TYPICAL WEIGHT CAPACITY: 6 - 22 pounds

Photograph provided by Hobart Brothers Company

COMPANY NAME: REIS MACHINES
MAJOR BUSINESS: Machinery
COUNTRY OF ORIGIN: West Germany
LERT: LR³ T
BASIC POWER SUPPLY: Electric
PRIMARY USES: Machine loading,
                        material handling
TYPICAL COST: $60,000
TYPICAL ACCURACY: ±.02 inch
TYPICAL WEIGHT CAPACITY: 55 pounds

Photograph provided by Reis Machines

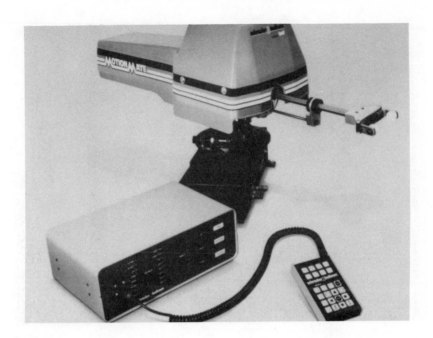

COMPANY NAME: SCHRADER BELLOWS
MAJOR BUSINESS: Industrial components
COUNTRY OF ORIGIN: USA
LERT: $L^3 R^2$
BASIC POWER SUPPLY: Pneumatic
PRIMARY USES: Material handling,
                                 machine loading
TYPICAL COST: $12,000
TYPICAL ACCURACY: ± .005 inch
TYPICAL WEIGHT CAPACITY: 5 pounds

Photograph provided by Schrader Bellows Division

COMPANY NAME: SEIKO INSTRUMENTS
MAJOR BUSINESS: Electrical equipment
COUNTRY OF ORIGIN: Japan
LERT: LR^3 T
BASIC POWER SUPPLY: Pneumatic
PRIMARY USES: Material handling
TYPICAL COST: $5,000 - 45,000
TYPICAL ACCURACY: ±.0004 inch to ±.001 inch
TYPICAL WEIGHT CAPACITY: 1 to 5 pounds

Photograph provided by Seiko Instruments U.S.A. Inc.

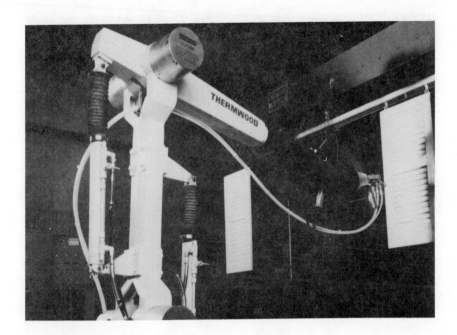

COMPANY NAME: THERMWOOD
MAJOR BUSINESS: Robots
COUNTRY OF ORIGIN: USA
LERT: $R^4 T$
BASIC POWER SUPPLY: Hydraulic
PRIMARY USES: Spray painting, assembly
TYPICAL COST: $30,000 - $50,000
TYPICAL ACCURACY: $\pm .010$ inch to $\pm .125$ inch
TYPICAL WEIGHT CAPACITY: 15 to 50 pounds

Photograph provided by Thermwood Corporation

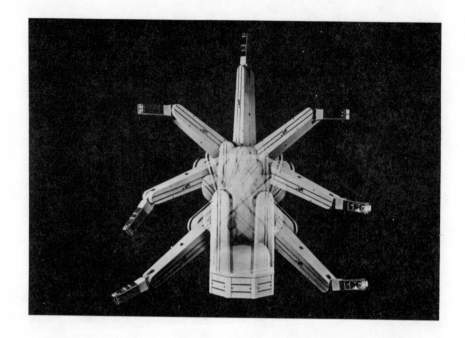

COMPANY NAME: UNITED TECHNOLOGIES
MAJOR BUSINESS: Aircraft, building products
COUNTRY OF ORIGIN: USA/Germany
LERT: R^4 T, R^5 T, L^3 R^3
BASIC POWER SUPPLY: Electric
PRIMARY USES: Spot weld, arc weld,
                             machine loading
TYPICAL COST: $80,000 - $100,000
TYPICAL ACCURACY: ±.010 inch
TYPICAL WEIGHT CAPACITY: 30 - 200 pounds (depends on
                             number of axes)

Photograph provided by United Technologies Automotive Group

COMPANY NAME: BINKS (THERMWOOD)
MAJOR BUSINESS: Spray painting equipment
COUNTRY OF ORIGIN: USA
LERT: R⁴ T
BASIC POWER SUPPLY: Electric
PRIMARY USES: Spray painting
TYPICAL COST: $45,000
TYPICAL ACCURACY: ±.125 inch
TYPICAL WEIGHT CAPACITY: 18 pounds

COMPANY NAME: GENERAL ELECTRIC
MAJOR BUSINESS: Electrical products
COUNTRY OF ORIGIN: Germany/Italy/Japan/USA
LERT: R⁴ T, L³ R²
BASIC POWER SUPPLY: Electric
PRIMARY USES: Arc weld, machine loading,
                          assembly, spot weld,
                          material handling
TYPICAL COST: Inquire
TYPICAL ACCURACY: ±.001 to ±.100 inch
TYPICAL WEIGHT CAPACITY: 14 to 132 pounds

COMPANY NAME: GMF (owned equally by
                          FANUC and GENERAL
                          MOTORS)
MAJOR BUSINESS: Robots
COUNTRY OF ORIGIN: Japan/USA
LERT: L³ R³, R⁴ T
BASIC POWER SUPPLY: Electric
PRIMARY USES: Arc weld, machine loading,
                          material handling,
                          spray painting
TYPICAL COST: Inquire
TYPICAL ACCURACY: ±.002 to ±.040 inch
TYPICAL WEIGHT CAPACITY: 20 to 175 pounds

COMPANY NAME: IBM
MAJOR BUSINESS: Computers
COUNTRY OF ORIGIN: USA/Japan
LERT: $L^3 R^2$
BASIC POWER SUPPLY: Electric
PRIMARY USES: Inspection, assembly
TYPICAL COST: $28,000 - $100,000
TYPICAL ACCURACY: $\pm$.002 to $\pm$.008 inch
TYPICAL WEIGHT CAPACITY: 5 - 15 pounds

COMPANY NAME: INTERNATIONAL
              ROBOMATION
              INTELLIGENCE
MAJOR BUSINESS: Robots
COUNTRY OF ORIGIN: USA
LERT: $R^4 T$
BASIC POWER SUPPLY: Air
PRIMARY USES: Material handling,
              machine loading
TYPICAL COST: $9,800
TYPICAL ACCURACY: $\pm$.040 inch
TYPICAL WEIGHT CAPACITY: 50 pounds

COMPANY NAME: PRAB
MAJOR BUSINESS: Robots
COUNTRY OF ORIGIN: USA
LERT: $L^3 R^2$
BASIC POWER SUPPLY: Hydraulic
PRIMARY USES: Press loading, material handling
TYPICAL COST: $30,000 - $150,000
TYPICAL ACCURACY: $\pm$.050 inch
TYPICAL WEIGHT CAPACITY: 50 to 2,000 pounds

COMPANY NAME: UNIMATE/WESTINGHOUSE
MAJOR BUSINESS: Electric products/appliances
COUNTRY OF ORIGIN: USA
LERT: $R^4\ T$
BASIC POWER SUPPLY: Hydraulic/electric
PRIMARY USES: Arc weld, spot weld,
material handling, forging,
die casting
TYPICAL COST: $27,000 - $80,000
TYPICAL ACCURACY: ±.008 inch, electric/±.080
inch, hydraulic
TYPICAL WEIGHT CAPACITY: 2 to 450 pounds